W0018249

Biological and Biomimetic Adhesives: Challenges and Opportunities

Biological and Biomimetic Adhesives

Challenges and Opportunities

Edited by

Romana Santos
University of Lisbon, Portugal
Email: romana.santos@campus.ul.pt

Nick Aldred
Newcastle University, UK
Email: nicholas.aldred@ncl.ac.uk

Stanislav Gorb
University of Kiel, Germany
Email: sgorb@zoologie.uni-kiel.de

Patrick Flammang
University of Mons, Belgium
Email: patrick.flammang@umons.ac.be

RSCPublishing

Publication of the Cost Action Group TD0906.

This volume is based on the proceedings of the 1st International Conference on Biological and Biomimetic Adhesives that was held at the University of Lisbon, Portugal 9-11 May 2012.

ISBN: 978-1-84973-669-5

A catalogue record for this book is available from the British Library

© COST Office 2013

All rights reserved

Apart from any fair dealing for the purpose of research or private study for non-commercial purposes, or criticism or review as permitted under the terms of the UK Copyright, Designs and Patents Act, 1988 and the Copyright and Related Rights Regulations 2003, this publication may not be reproduced, stored or transmitted, in any form or by any means, without the prior permission in writing of The Royal Society of Chemistry or the copyright owner, or in the case of reprographic reproduction only in accordance with the terms of the licences issued by the Copyright Licensing Agency in the UK, or in accordance with the terms of the licences issued by the appropriate Reproduction Rights Organization outside the UK. Enquiries concerning reproduction outside the terms stated here should be sent to The Royal Society of Chemistry at the address printed on this page.

The RSC is not responsible for individual opinions expressed in this work.

Published by The Royal Society of Chemistry,
Thomas Graham House, Science Park, Milton Road,
Cambridge CB4 0WF, UK

Registered Charity Number 207890

Visit our website at www.rsc.org/books

Printed in the United Kingdom by CPI Group (UK) Ltd, Croydon, CR0 4YY

EUROPEAN
SCIENCE
FOUNDATION

ESF provides the COST Office through an EC contract

COST is supported by the EU RTD Framework programme

EUROPEAN COOPERATION
IN SCIENCE AND TECHNOLOGY

COST – the acronym for European Cooperation in Science and Technology- is the oldest and widest European intergovernmental network for cooperation in research. Established by the Ministerial Conference in November 1971, COST is presently used by the scientific communities of 36 European countries to cooperate in common research projects supported by national funds.

The funds provided by COST - less than 1% of the total value of the projects - support the COST cooperation networks (COST Actions) through which, with EUR 30 million per year, more than 30 000 European scientists are involved in research having a total value which exceeds EUR 2 billion per year. This is the financial worth of the European added value which COST achieves.

A "bottom up approach" (the initiative of launching a COST Action comes from the European scientists themselves), "à la carte participation" (only countries interested in the Action participate), "equality of access" (participation is open also to the scientific communities of countries not belonging to the European Union) and "flexible structure" (easy implementation and light management of the research initiatives) are the main characteristics of COST.

As precursor of advanced multidisciplinary research COST has a very important role for the realisation of the European Research Area (ERA) anticipating and complementing the activities of the Framework Programmes, constituting a "bridge" towards the scientific communities of emerging countries, increasing the mobility of researchers across Europe and fostering the establishment of "Networks of Excellence" in many key scientific domains such as: Biomedicine and Molecular Biosciences; Food and Agriculture; Forests, their Products and Services; Materials, Physical and Nanosciences; Chemistry and Molecular Sciences and Technologies; Earth System Science and Environmental Management; Information and Communication Technologies; Transport and Urban Development; Individuals, Societies, Cultures and Health. It covers basic and more applied research and also addresses issues of pre-normative nature or of societal importance. For more information the web address is: http://www.cost.eu

This publication is supported by COST.

PREFACE

A. del Campo[1], W. Schwotzer[2], S.N. Gorb[3], N. Aldred[4], R. Santos[5] and P. Flammang[6]

[1] Max-Planck-Institut für Polymerforschung, Mainz, Germany
[2] nolax AG, Sempach Station, Switzerland
[3] Functional Morphology and Biomechanics, University of Kiel, Germany
[4] School of Marine Science and Technology, Newcastle University, UK
[5] Biomedical and Oral Sciences Research Unit, University of Lisbon, Portugal
[6] Biology of Marine Organisms and Biomimetics, University of Mons, Belgium

1 INTRODUCTION

For decades, nature's adhesion and bonding techniques have fascinated laypeople and professionals alike. Fundamentally, all living things are glued together, with most of their components connected by adhesive bonds. There are many organisms for which adhesion is vital in conjunction with food procurement, locomotion, attachment and defence, particularly among primeval life forms. Frequently cited examples include the prey-trapping structures of some carnivorous plants, the adhesion organs of insects and reptiles, the cements of mussels and barnacles, and the mucus of amphibians and snails.

Many natural bioadhesives perform in ways that man-made products simply cannot match. Some are reversible, others work most effectively underwater and many are universal in their performance to substrates of varying composition and structure. For these reasons, it is not surprising that reports on adhesion in nature rank among the most regularly recurring topics in the science sections of newspapers and popular science journals. Such articles rarely omit the prediction that bio-inspired adhesives will soon revolutionize medical joining technologies and totally supplant classic surgical sutures, staples, or screws. In the real world, however, things look a little different. In reality the deployment of biomimetic systems is the exception rather than the rule.[1] The purpose of COST Action TD0906 "Biological Adhesives – From Biology to Biomimetics", which was launched in 2010, is to draw together expertise bridging every discipline that will be required to contribute to successful development of bio-inspired adhesives, and to ask the question why, at present, is there such disparity between vision and reality?

2 ADHESIVE STRATEGIES IN NATURE

From the biologists' point of view, natural adhesives serve a variety of functions, including: (1) temporary attachment of body parts together; (2) attachment of one organism to another (copulation, phoresy or parasitism, prey capture); or (3) attachment of an organism to a non-living surface, including dynamic attachment during locomotion and permanent fixation. The evolutionary background and the biology of species on the one hand, and environmental constraints on the other hand, both influence the specific design

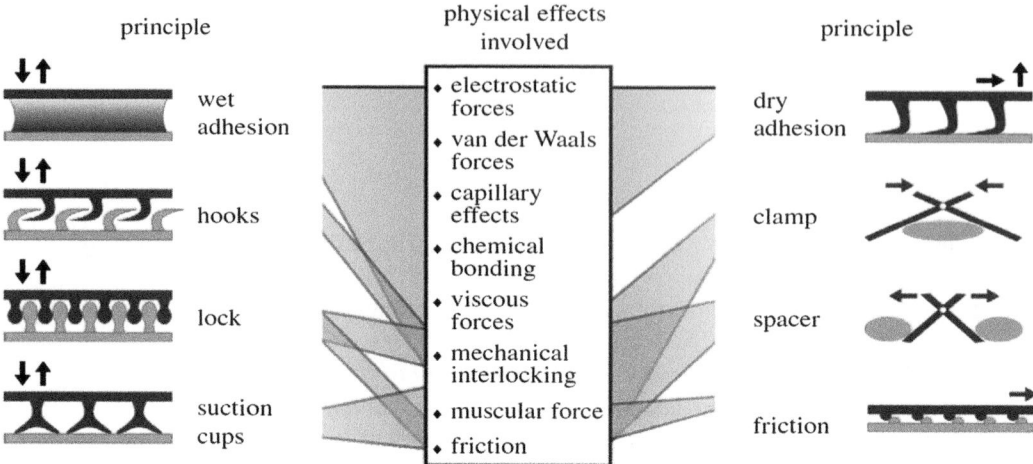

Figure 1 *Functional principles of biological attachment structures and physical effects involved.[2]*

of adhesion mechanisms in a particular organism. The diversity of biological attachment devices is therefore huge, although some common principles have evolved independently in different biological lineages.

Biological attachment systems can be subdivided into several groups according to the following principles: (1) fundamental physical mechanisms, according to which the system operates, (2) biological function of the attachment device, and (3) duration of the contact. Eight fundamental attachment mechanisms have been previously recognized: (1) hooks, (2) lock or snap, (3) clamp, (4) spacer or expansion anchor (5) suction, (6) dry adhesion, (7) wet adhesion (glue/cement, capillarity), and (8) friction (Figure 1).[2] However, various combinations of these principles may also occur in real biological systems.

Terrestrial animals that climb on plants, rocks or other types of unpredictable substrates have evolved adhesive organs on their feet of two basic designs: (1) pads densely covered with specialised μm-sized setae, and (2) pads with a relatively smooth surface profile. Some of these attachment pads are supplemented with various kinds of fluids (reversible wet adhesives), and some not (dry adhesives). Flies, beetles, large spiders, and gekkonid lizards, have feet possessing adhesive organs of the hairy type (Figure 2), while tree frogs, bees, and grasshoppers have adhesive organs of the smooth type. Both mechanisms allow reversible adhesion by maximising contact with the substrate, almost regardless of its microsculpture. In addition, they are reusable, enabling the animal to attach and detach its feet repeatedly, self-cleaning and wear-resistant. Detachment occurs at will and with negligible force. In addition, they present a non-sticky default state so that adhesion only occurs when it is required, i.e. on-demand. All of these features are required for climbing animals and are desirable properties for novel synthetic adhesives. Of the two types of structures, hairy pads of geckos and insects (Figure 2) have been studied most extensively, and have become a model system for bioinspired adhesive research.[3]

The situation is rather different in the aquatic environment. Shear forces underwater are much stronger than on land. They are also directionally unpredictable, requiring adhesion systems that are sturdy in all directions. That is why adhesion is "a way of life in the sea".[4] Marine bacteria, algae and animals produce adhesive secretions that are able to

Figure 2 *Examples of dry (left) and and wet (right) reversible adhesive designs in nature. Left: Gecko feet. Right: Foot of the chrysomelid beetle Gastrophysa viridula.*

bond surfaces tenaciously underwater. Attachment devices developed by marine organisms therefore tend to rely on highly viscous or solid adhesive secretions, mostly containing adhesive proteins. The diversity of marine adhesives is huge but, to date, only a very limited number of marine organisms have been studied in detail and most of this diversity remains unexplored.

Three types of marine adhesives may be distinguished depending on their mode of operation and their composition. Permanent adhesion involves the secretion of an adhesive that hardens with time and is characteristic of sessile organisms that remain in the same place throughout their life. From a molecular point of view, most studies have focused on the permanent adhesives from mussels and barnacles (Figure 3a) because these organisms are important macro-foulers of marine structures and the characterization of their adhesives originally aimed towards development of improved antifouling strategies. Many marine organisms, however, use temporary adhesion at one or more stages in their life cycle. Temporary adhesion allows simultaneous adhesion and locomotion, thus allowing adult organisms to graze, hunt or locate a mate, and larval forms to explore immersed surfaces prior to permanent adhesion. Some organisms such as gastropod molluscs attach by a viscous film they produce between their body and the substratum, creeping on this film which is left behind them as they move. Others, such as echinoderms, attach firmly but only very briefly to the substratum. Their adhesive organs, the so-called tube feet (Figure 3b), enclose a duo-gland system comprising two types of secretory cells: cells releasing an adhesive material and cells producing a de-adhesive secretion. Using these antagonistic secretions, tube feet can attach and detach repetitively. Finally, instantaneous adhesion is characterized by an explosive release of adhesive, destroying the adhesive organ which may be used only once. The peculiar Cuvieran tubules of sea cucumbers are typical instantaneous adhesive organs. These tubules are expelled as white threads that function as a defence mechanism against predators (Figure 3c). Once in seawater, they elongate significantly and instantly become sticky upon contact.

Functional convergences are frequently noted among marine animals in terms of the type and mechanisms of adhesion used, but molecular convergences have also been identified. It is now well-established that the common characteristics of different marine adhesives (e.g., ability to displace water from the substratum, to spread and rapidly form

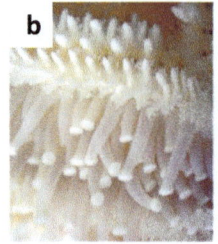

Figure 3 *Marine adhesive systems a) Two top biological models together: barnacles growing on mussels. The white conical barnacles are cemented to the shell of a mussel. The bivalve, in turn, is attached to other mussels with byssal threads. These two organisms are the most intensely studied in terms of adhesive characterization. b) Sea stars can attach strongly to a variety of underwater substrata using a multitude of small appendages, the so-called tube feet. These organs are made up of two parts: a proximal extensible stem which allows movement of the tube foot, and a distal flat-shaped disc which is responsible for attachment to the surface. c) Sea cucumbers expel sticky spaghetti-like tubules to trap attacking organisms.*

strong adhesive bonds with the surface) are often related to post-translational modifications of the adhesive proteins, such as hydroxylation, phosphorylation and glycosylation; an exception being the case of the barnacle where there is presently no evidence for post-translational modification of adhesive proteins. Hydroxylation is the most thoroughly investigated modification at present. Studies on mussels and a few other marine animals have revealed the high content of the unusual amino acid 3,4-dihydroxy-L-phenylalanine (DOPA) in the adhesive composition, which is formed by post-translational modification of tyrosine.[5] This modified amino acid plays important interfacial and cross-linking roles in marine adhesive secretions.

3 PROGRESS IN BIO-INSPIRED ADHESIVE RESEARCH

The scientific community studying bioinspired adhesion includes zoologists, bio- and polymer-chemists, physicists and micro- and nanofabrication technologists, working in a multidisciplinary environment to study, theoretically analyze, and prototype mechanisms and principles of natural adhesive models. Important progress has been accomplished in the last ten years, with most reported works focusing on 'gecko-inspired' reversible dry adhesives[6] and 'mussel-inspired' permanent glues for underwater applications.[5,7]

In contrast to conventional adhesives, the design of gecko-inspired adhesives does not rely on complex reactive formulations, but on the combination of micro and nanofabrication approaches to obtain a hierarchical topographical design (Figure 4). Fibrillar surfaces with increasing complexity have been reported.[6] First attempts to obtain densely packed high aspect ratio nanofibrils for maximizing adhesion performance were followed by tilted and asymmetric fibrillar designs to obtain directional and reversible adhesion (Figure 4a-b) and hierarchical arrangements for better adaptability to rough surfaces (Figure 4c). More recently patterning technologies were combined with responsive polymer materials to realize actuated fibrils, enabling adhesion on-demand by application of external stimuli like temperature or magnetic fields (Figure 4d-e). A

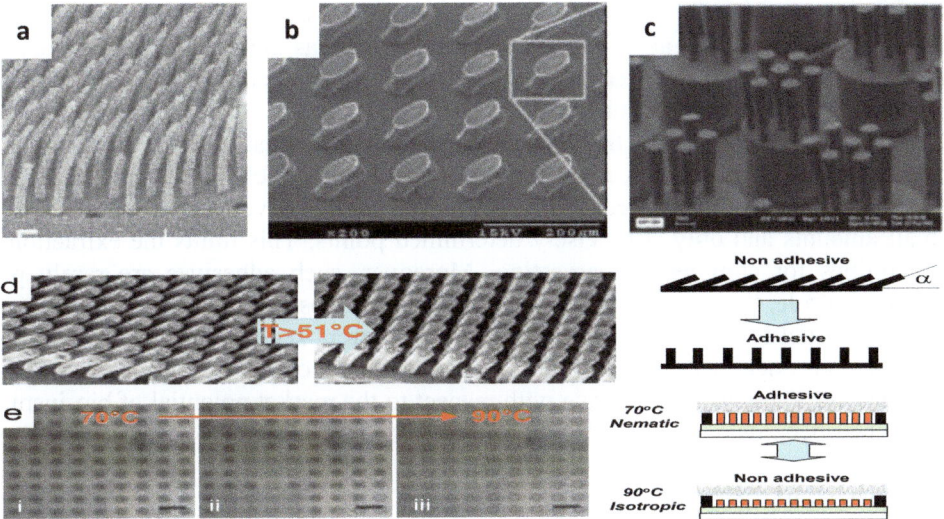

Figure 4 *Gecko-inspired adhesive surfaces showing asymmetric and tilted configurations of fibrils[8,9] (a,b) and hierarchical organization[9] (c). Reversible adhesive concepts based on responsive pillar patterns of shape memory polymers that change tilting direction[11] (d) or of liquid crystalline elastomers which change elongation[12] (e) with temperature.*

valuable collection of fabrication strategies and gecko-inspired working systems is now demonstrated at laboratory level and the first manufacturing concepts and products are arising.

Research into mussel-inspired adhesives has developed broadly in two different directions: the production of recombinant mussel adhesive proteins and the incorporation of DOPA related units to polymeric chains. The presence of DOPA increases the affinity of the adhesive formulation for oxide surfaces in the presence of water and, consequently, achieves a higher interfacial strength. As an additional advantage, the catechol unit of DOPA mediates self-healing properties when incorporated inside the adhesive network. Most mussel-inspired adhesives developed within the last 5 years have focused on biotechnological and biomedical applications. DOPA-containing hydrogels have been suggested as adhesives for surgery because of their biocompatibilty and their ability to attach, react and self-polymerize *in-vivo*.[7] Recombinant adhesive proteins have been recently entered the market as adhesive coatings for cell culture.[13]

Recent reports have also applied the adhesive mechanism of the sandcastle worm to biomedical materials. In this case a mixture of highly charged proteins forms an insoluble and sticky complex coacervate that further solidifies via DOPA crosslinking. Synthetic mimics using polycations and polyanions have been realized.[14]

4 BIOLOGICAL AND BIOMIMETIC ADHESIVES IN PRACTICE

4.1 Practicability Requirements

To be practicable, an adhesive system must fulfil a wide range of criteria. While the technical adhesion profile plays a significant role, it does not constitute a guarantee for commercial success. Equally important criteria include the availability of raw materials, reproducibility in production and consistent quality, as well as the price/performance ratio in comparison with alternative technologies. In medical applications, toxicological and allergy-relevant aspects must be considered as well.

Likely the biggest obstacle that a biological adhesive must overcome en route to commercialization is processing cost. This cost is comparatively high simply because nature is extremely resource-efficient: high-performance adhesives are produced only in very small amounts and only at precisely determined points. This limits the extraction of larger quantities for commercial harvesting. Moreover, such adhesives are rarely pure substances; as a rule, they are complex reactive systems whose components are not mixed by the organism until they are applied. So, given the current state of the art in the field of biotechnology, production via recombinant DNA would be prohibitively complex, which is to say expensive. There is a simple rule with respect to the market potential of bio-inspired adhesives: their benefits must justify their inevitably higher cost.

4.2 Copy Instead of Original

Biomimetic, or bio-inspired, adhesives do not necessarily have to be composed of complex biopolymers. It may, in fact, be more advantageous to merely glean the active components and functional principles from nature and approach these through synthetic means. It is important to note that the focus here is on *adhesion* and that the desired *cohesive* properties of an adhesive material can often be achieved in artificial systems using the resources presently available. In terms of *adhesion*, there are three salient properties in biological systems that are difficult to emulate in synthetic equivalents:

- Reversibility and response, i.e. bonding/debonding on demand
- Bonding underwater or on wet substrates
- Adhesion to substrates with variable physical/chemical surface properties

Today, a number of commercially utilized joining systems can already be legitimately referred to as biomimetic. Probably the best-known example is Velcro tape, in which microscopically small hooks get entangled in even tinier loops. A patent application was filed for this principle in 1951 and the patent was granted in 1954 (Swiss patent No. 2956389). More recently, anticipation of a practical gecko-inspired adhesive has been growing and the potential market is considerable; presuming the demand currently served by pressure sensitive adhesives which have a global market volume of ~25 billion square meters per year.[15]

Occasionally, biological bonding structures combine principles of mechanical interlocking, van der Waals interactions, chemical bonds, viscous and capillary forces. This does not imply that these principles evolved simultaneously, but it does suggest that various combinations of these principles resulted in an evolutionary benefit for the respective organism. This approach is also being considered for commercial purposes (WO 2008 091 386; Table 1).

4.3 Patents: Pointers to Planned Commercial Applications

The path from scientific insight to commercial feasibility is usually long and costly. It is therefore littered with patent applications and even granted patents. The intention, of course, is to secure a certain degree of market sovereignty. Thus, the perusal of relevant patent literature is a good way to identify technological trends. However, there are two caveats: one is generic, the other relates to the topic of 'bionics'. In general, it must be considered that the disclosure of a patent application does not occur until at least 18 months after it was submitted, so the last one and a half years are obscure. Further, terms like "bionics", "biomimetics", and others are neither protected nor clearly defined in the patent vocabulary. Therefore, the presence of these terms in patent applications does not necessarily imply that the concept on which an invention is based was inspired by observations of natural phenomena, or vice versa. Nonetheless, the study of patent literature is extremely useful because it not only describes the state of the art, but also points to unmet market needs.

The so-called theory of inventive problem solving is based upon three main principles in engineering: (1) all patents are based on about 40 inventive principles; (2) all technology trends are predictable; (3) important inventions come from outside the industry within which they are applied. Throughout evolution, nature has constantly been called upon to act as an engineer in solving adhesion problems. Keeping this in mind an engineer can extend his pool of ideas by looking at biological adhesion principles.

4.4 Technology Trends in Biomimetic Adhesives

Table 1 lists a selection of recent patents that involve "biomimetic adhesives". It does not purport to be complete in the sense of an exhaustive patent search. Rather, a comparison of the conclusions stated in the patents suggests the following trends for the possible use of biomimetic adhesive systems:

Dry reversible adhesion: The advantage of 'gecko-inspired' adhesive surfaces is that they are not tacky, do not require any adhesive fluid and can be separated again without leaving any residue. A host of conceivable practical applications exists for such adhesive films. They can be useful in the development of mobile robots that rely on them to move in three-dimensional space. Additionally, they provide new options for the design of novel manipulators and handling devices, and provide some potential for medical applications in dermatology and surgery. Potentially, they can replace pressure-sensitive adhesives in some specific applications.

Underwater bonding: The bond profiles of adhesives used by marine organisms are presently unmatched by commercial adhesive systems. Dihydroxyphenyl radicals can enhance adhesive properties in two ways: as a coupling agent and, when oxidized, as a crosslinker. The comparatively large number of patents devoted to this domain justifies the assumption that considerable activity is ongoing in the field, and that such adhesive systems can be expected in the marketplace in the near future.

Dry/wet systems: The combination of specially structured surfaces and specific fluids, such as mucus (tree frog pads), lipids (insect pads) or protein-lipid mixtures observed in various species also characterizes biomimetic systems.

Antifouling: At first sight, one could surmise that antifouling coatings are the antithesis of adhesives. However, it should be considered that such coatings would ideally be anisotropic in their adhesive potential, adhering to the coated substrate more effectively than to the unwanted fouling organisms. It appears that the equivalent adhesion principles of mussel proteins are at the focus of R&D strategies for such applications as well.

Table 1 *Patent activities involving biomimetic adhesives*

Patent number	Application	Functional principle
WO2003 0087 338	adhesive profile in the style of MAP (Mussel Adhesive Protein)	DOPA/catechol residues attached to synthetic polypropylene/polyethylene oxide block-polymers
US2006 005 362	fasteners based on dry adhesion	nano-structured surface
US2006 0029 997	foot adhesive protein of Mytilus edulis (Mussel)	cloning and expression of recombinant DNA
EP 2 094 123	fasteners for textiles and foils based on dry adhesion	nano-structured surface
WO2008 091 386	adhesion in the style of biological systems	nanostructured surface combined with a biomimetic adhesive mass containing dihydroxyphenyl residues
US2009 0093 610	anchoring of thin layers on surfaces	catechol-functionalized polymers
EP 1 960 849	waterproof acrylic adhesive	DOPA-functionalized poly-acrylate
US2010 0137 903	carrier material coated with adhesive for medicinal purposes	DHP (dihydroxyphenyl) modified backbone polymer
US2011 0015 759	adhesives for medicinal applications	polyphenols in combination with water soluble polymers such as polysaccharides
US2011 0076 504	multifunctional antifouling coating	DHP functionalized poly-electrolyte
US2011 0123 477	antifouling coating	In-situ generated polymer in combination with reactive groups such as noradrenalin, DOPA, etc.
US2011 0171 239	gel from silk protein for medicinal use	silk fibroin
US2011 0105 712	antifouling coating	zwitterionic polymer with PHP function
US2011 1173 321	dry adhesion	nano-structured surface from elastomers
CN 102 153 867	robot footpads	setae made of silicon rubber
CN 101 774 528	dry adhesion	nano-structured surface
US2012 0016 390	adhesives for medicinal applications antifouling coating	dendrimers modified with DHP
US2012 0003 888	carrier material coated with adhesive for medicinal purposes	DOPA modified PEG (polyethylene glycol)
US2012 0078 296	improved adhesion of adhesives and sealants	DHP modified backbone polymer

Hydrogels for medical applications: The first phase in a biological adhesion processes often involves the secretion of a gel-like substance into the adhesive joint. Subsequently, the gel matrix is often crosslinked to create a durable bond. Protein gelling agents (such as silk fibroin) or polysaccharides are recommended as thickeners for medical applications. Sometimes the crosslinked substances within the joint form a foam-like material, which prevents crack propagation through the adhesive joint.

Mechanisms for applying minute quantities of glue: Biological adhesives provide interesting ideas for the design of microdispensive systems capable of applying small droplets of glues to the surface or rapid generation of ultrathin adhesive films on the surface.

5 PROCEEDINGS OUTLOOK

The above-mentioned themes (reversibility, underwater adhesion, dry bonds, etc.) are far from being accomplished on an industrial scale. On the other hand, fundamental research in the field of biological and biomimetic bonding has made enormous progress in recent years. The improved scientific collaboration and interdisciplinary cooperation inspired by COST Action TD0906 were exemplified by the 1st International Conference on Biological and Biomimetic Adhesives, held in Lisbon in May 2012, and a cross-section of the key themes addressed at this event are presented in this volume. Representative activities in the areas of bioadhesive characterisation [Part 1], modelling of biomimetic systems [Part 2], targeting specific applications [Part 3] and surface modification for optimal bonding/debonding [Part 4] were all covered by specialists in their respective fields.

The first question that emerges in the examination of adhesion phenomena relates to the underlying mechanisms. Are the bonds adhesive, friction-fit, form-locking, or a combination of these? If an adhesive bond is involved, considerable attention has to be devoted to the chemical nature of the adhesive. The complexity of biological polymers and the often-minimal substance quantities available for analysis pose formidable challenges for chemical and biochemical analyses. The key issues are to understand the framework structures, identify the functional groups that participate in the adhesion process and elucidate their spatial configurations. The biosynthesis of these substances and the responsible cellular structures are also of interest from the biological standpoint. Contributions to this book draw on poorly-known biological adhesion mechanisms from e.g. bacteria [Cuesta-Garrote et al.], algae [Dimartino et al.] and slugs [Smith].

Many biological adhesives involve structures of micro- and nanoscopic dimensions. Accordingly, the local forces are minimal as well. It is challenging for experimental technologists to measure and model such interactions at the local and global scale. Conversely, examinations of this kind afford in-depth insights into adhesive bonding phenomena, so they are of great interest not only in the context of biological systems but also of direct practical benefit. The very latest methods for characterizing surfaces have been applied to the natural or biomimetic adhesive systems of starfish [Higgins and Mostaert], mussels [Birkedal et al.] and octopus [Tramacere et al.]. The structural data obtained can then be used to refine theoretical models as for gecko-like adhesives [Röhrig et al.].

Ultimately, the objective must be to leverage the insights gained from academic research to develop viable commercial products. An innovative thrust is needed to factor cost-effectiveness into the equation. As discussed earlier, it will be technically and economically difficult to utilize natural materials directly and the more viable approach will be to emulate the functionality of biological systems with the methods offered by materials science. Only in this way will it be possible to develop highly specialized adhesives for challenging applications with a truly interesting price/performance ratio. A number of contributions to this book take an important step further, bridging the gap that currently exists between characterisation, synthesis and application [Röhrig et al.; Kizilkan et al.; Chung and Chaudhury; Tramacere et al.; Ceylan et al.; Dimartino et al.; Cuesta-Garrote et al.; Zietek et al.; and Trzaskowski et al.] and the rewards for this endeavour may be considerable.

It is clear that a renaissance in our approach to developing novel bioadhesive materials is underway, and we can certainly expect new commercial bonding technologies to emerge within the foreseeable future.

Acknowledgements

Parts of this preface (including figures) have been reprinted from references 16 and 17. We thank Springer for granting us the permission to re-use this material.

References

1   A.P. Duarte, J.F. Coelho, J.C. Bordado, M.T. Cidade and M.H. Gil, *Prog. Polymer Sci.* 2012, **37**, 1031.

2   S.N. Gorb, *Phil. Trans. R. Soc A,* 2008, **366**, 1557.

3   S.N. Gorb, M. Sinha, A. Peressadko, K.A. Daltorio and R.D. Quinn, *Bioinspir. Biomim.* 2007, **2**, S117.

4   J.H.Waite, *Biol. Rev.*1983, 58: 209.

5   B.P. Lee, P.B. Messersmith, J.N. Israelachvili and J.H. Waite, *Ann. Rev. Mater. Res.* 2011, **41**, 99.

6   L.F. Boesel, C. Greiner, E. Arzt and A. del Campo, *Adv. Mater.* 2010, **22**, 2125.

7   C.E. Brubaker and P.B. Messersmith, *Langmuir* 2012, **28**, 2200.

8   T.I. Kim, H.E. Jeong, K.Y. Suh and H.H. Lee, *Adv. Mater.* 2009, **21**, 2276.

9   M.P. Murphy, B. Aksak and M. Sitti, *Small* 2009, **5**, 170.

10   C. Greiner, E. Arzt and A. del Campo, *Adv. Mater.* 2009, **21**, 479.

11   S. Reddy, E. Arzt and A. del Campo, *Adv. Mater.* 2007, **19**, 3833.

12   J. Cui, D. M. Drotlef, I. Larraza, J. P. Fernández-Blázquez, L. F. Boesel, C. Ohm, M. Mezger, R. Zentel and A. del Campo, *Adv. Mater.* 2012, **24**, 4601.

13   *MAPtrix, www.kollodis.com.*

14   H. Shao and R.J. Stewart, *Adv. Mater.* 2010, **22**, 729.

15   Z. Czech and R. Milker, *Mater. Sci-Poland* 2005, **23**, 1015.

16   A. Del Campo, W. Schwotzer, S. Gorb and Flammang P. Adhäsion Kleben und Dichten, 2012, **56(9)**, 37.

17   W. Schwotzer, A. Del Campo, S. Gorb and Flammang P. Adhäsion Kleben und Dichten, 2012, **56(10)**, 38.

Contents

List of Contributors

N. Aldred
School of Marine Science and Technology
Newcastle University
Newcastle upon Tyne NE1 7RU, UK
E-mail: nicholas.aldred@ncl.ac.uk

M. Almeida
Unidade de Investigação em Ciências Orais e
Biomédicas (UICOB)
Faculdade de Medicina Dentária da Universidade
de Lisboa
Cidade Universitária
1649-003 Lisboa, Portugal

F. Arán-Ais
Footwear Technological Institute-INESCOP
Polígono Industrial Campo Alto
03600 Elda (Alicante), Spain

R. Azhari
Department of Biotechnology Engineering
ORT Braude College
PO Box 78, Karmiel 21982, Israel

L. Beccai
Center for Micro-BioRobotics@SSSA
Istituto Italiano di Tecnologia
Viale Rinaldo Piaggio 34
56025 Pontedera, Italy
E-mail: lucia.beccai@iit.it

H. Birkedal
iNANO and Department of Chemistry
Aarhus University
Gustav Wieds Vej 14
DK-8000 Aarhus C, Denmark
E-mail: hbirkedal@chem.au.dk

S. Bundschuh
Institute for Applied Materials (IAM)
Karlsruhe Institute of Technology (KIT)
Hermann-von-Helmholtz-Platz 1
76344 Eggenstein-Leopoldshafen, Germany

H. Ceylan
Institute of Materials Science and
Nanotechnology
National Nanotechnology Research Center
(UNAM)
Bilkent University
Ankara 06800, Turkey

M.K. Chaudhury
Lehigh University
Department of Chemical Engineering
111 Research Drive
Bethlehem, 18015, Pennsylvania, USA

J.Y. Chung
Lehigh University
Department of Chemical Engineering
111 Research Drive
Bethlehem, 18015, Pennsylvania, USA

and

Harvard University
School of Engineering and Applied Sciences
60 Oxford Street
Cambridge, 02138, Massachusetts, USA
E-mail: jchung@seas.harvard.edu

T. Ciach
Biomedical Engineering Laboratory
Faculty of Chemical and Process Engineering
Warsaw University of Technology
Warynskiego 1
00-645 Warsaw, Poland

N. Cuesta-Garrote
Footwear Technological Institute-INESCOP
Polígono Industrial Campo Alto
03600 Elda (Alicante), Spain

A. del Campo
Max-Planck-Institut für Polymerforschung
Ackermannweg 10
55128 Mainz, Germany
Email: delcampo@mpip-mainz.mpg.de

S. Dimartino
Biomolecular Interaction Centre
University of Canterbury
Private Bag 4800
Christchurch 8140, New Zealand
E-mail: simone.dimartino@canterbury.ac.nz

M.J. Escoto-Palacios
Footwear Technological Institute-INESCOP
Polígono Industrial Campo Alto
03600 Elda (Alicante), Spain
E-mail: mjescoto@inescop.es

P. Flammang
University of Mons - UMONS
Biology of Marine Organisms and Biomimetics
20 Place du Parc
7000 Mons, Belgium
E-mail : Patrick.Flammang@umons.ac.be

S. Frølich
iNANO and Department of Chemistry
Aarhus University
Gustav Wieds Vej 14
DK-8000 Aarhus C, Denmark

S.N. Gorb
Department of Functional Morphology and
Biomechanics
Zoological Institute at the University of Kiel
Am Botanischen Garten 1–9
24098 Kiel, Germany
E-mail: sgorb@zoologie.uni-kiel.de

M.O. Guler
Institute of Materials Science and
Nanotechnology
National Nanotechnology Research Center
Bilkent University
Ankara 06800, Turkey
E-mail: moguler@unam.bilkent.edu.tr

M. Haber
Biota Ltd.
652/8 Hanassi St.
Or-Akiva 30600, Israel

L. Heepe
Department of Functional Morphology and
Biomechanics
Zoological Institute at the University of Kiel
Am Botanischen Garten 1–9
24098 Kiel, Germany

L.J. Higgins
School of Biology and Environmental Science
University College Dublin
Belfield, Dublin 4, Ireland
E-mail: laila.higgins@ucdconnect.ie

H. Hölscher
Institute of Microstructure Technology (IMT)
Karlsruhe Institute of Technology (KIT)
Hermann-von-Helmholtz-Platz 1
76344 Eggenstein-Leopoldshafen, Germany

E. Kizilkan
Department of Functional Morphology and
Biomechanics
Zoological Institute at the University of Kiel
Am Botanischen Garten 1–9
24098 Kiel, Germany
E-mail: ekizilkan@zoologie.uni-kiel.de

H. Leemreize
iNANO and Department of Chemistry
Aarhus University
Gustav Wieds Vej 14
DK-8000 Aarhus C, Denmark

I. Lir
Biota Ltd.
652/8 Hanassi St.
Or-Akiva 30600, Israel

L.P. Lopes
Unidade de Investigação em Ciências Orais e
Biomédicas (UICOB)
Faculdade de Medicina Dentária da Universidade
de Lisboa
Cidade Universitária
1649-003 Lisboa, Portugal

M. Lopes
Unidade de Investigação em Ciências Orais e
Biomédicas (UICOB)
Faculdade de Medicina Dentária da Universidade
de Lisboa
Cidade Universitária
1649-003 Lisboa, Portugal

B. Mazzolai
Center for Micro-BioRobotics@SSSA
Istituto Italiano di Tecnologia
Viale Rinaldo Piaggio 34
56025 Pontedera, Italy
E-mail: barbara.mazzolai@iit.it

A.S. Mostaert
School of Biology and Environmental Science
University College Dublin
Belfield, Dublin 4, Ireland
E-mail: Anika.Mostaert@ucd.ie

and

Conway Institute of Biomolecular and
Biomedical Research
University College Dublin
Belfield, Dublin 4, Ireland

C. Orgilés-Barceló
Footwear Technological Institute-INESCOP
Polígono Industrial Campo Alto
03600 Elda (Alicante), Spain

M. Röhrig
Institute of Microstructure Technology (IMT)
Karlsruhe Institute of Technology (KIT)
Hermann-von-Helmholtz-Platz 1
76344 Eggenstein-Leopoldshafen, Germany
E-mail: michael.roehrig@kit.edu

R. Santos
Unidade de Investigação em Ciências Orais e
Biomédicas (UICOB)
Faculdade de Medicina Dentária da Universidade
de Lisboa
Cidade Universitária
1649-003 Lisboa, Portugal
Email: romana.santos@campus.ul.pt

W. Schwotzer
nolax AG
Sempach Station, Switzerland
E-mail: willi.schwotzer@nolax.com

A.M. Smith
Ithaca College
Department of Biology
953 Danby Road, Ithaca, NY 14850, USA
E-mail: asmith@ithaca.edu

R. Stallbohm
iNANO and Department of Chemistry
Aarhus University
Gustav Wieds Vej 14
DK-8000 Aarhus C, Denmark

A. B. Tekinay
Institute of Materials Science and
Nanotechnology
National Nanotechnology Research Center
Bilkent University
Ankara 06800, Turkey
E-mail: atekinay@unam.bilkent.edu.tr

M. Thiel
Nanoscribe GmbH
Hermann-von-Helmholtz-Platz 1
76344 Eggenstein-Leopoldshafen, Germany

F. Tramacere
Center for Micro-BioRobotics@SSSA
Istituto Italiano di Tecnologia
Viale Rinaldo Piaggio 34
56025 Pontedera, Italy
E-mail: francesca.tramacere@iit.it

and

BioRobotics Institute
Scuola Superiore Sant'Anna
Viale Rinaldo Piaggio 34
56025 Pontedera, Italy

M. Trzaskowski
Biomedical Engineering Laboratory
Faculty of Chemical and Process Engineering
Warsaw University of Technology
Warynskiego 1
00-645 Warsaw, Poland
E-mail: m.trzaskowski@ichip.pw.edu.pl

Y.H. Tseng
iNANO and Department of Chemistry
Aarhus University
Gustav Wieds Vej 14
DK-8000 Aarhus C, Denmark

M. Worgull
Institute of Microstructure Technology (IMT)
Karlsruhe Institute of Technology (KIT)
Hermann-von-Helmholtz-Platz 1
76344 Eggenstein-Leopoldshafen, Germany

P.A. Zietek
Faculty of Chemical and Process Engineering
Warsaw University of Technology
Warynskiego 1
00-145 Warsaw, Poland
E-mail: P.Zietek@ichip.pw.edu.pl

BIOADHESIVE CHARACTERISATION

MULTIPLE METAL-BASED CROSS-LINKS: PROTEIN OXIDATION AND METAL COORDINATION IN A BIOLOGICAL GLUE

A.M. Smith

Ithaca College, Department of Biology, Ithaca NY, USA
asmith@ithaca.edu

1 INTRODUCTION

Mucus hydrogels are ubiquitous materials that typically serve a lubricating function. These gels contain more than 95% water, and gain their integrity from tangling of unusually large polymers.[1] These polymers form a loose network that traps water and maintains shape, but is typically not designed for great mechanical strength. Nevertheless, a number of gastropod molluscs can use similar hydrogels to create strong adhesion.[2] Limpets use adhesive gels to create tenacities of 200-500 kPa.[1] The adhesive gel of the terrestrial slug *Arion subfuscus* can be used to glue discs together and resist stresses of up to 100 kPa.[2] The combination of adhesive performance with high water content is unusual, and likely to be highly useful in a medical context.

The primary structural features responsible for creating tough adhesives out of hydrogels are the presence of specific proteins and metals. While most forms of mucus consist primarily of large carbohydrate-rich polymers, adhesive hydrogels also have a significant protein content.[2] In all cases studied, there are specific proteins that are characteristic of the glue, and these proteins have gel-stiffening activity; they can be added to different commercial gel-forming polymers and cause up to a 50 fold increase in stiffness.[3] In addition, metals are present in these glues. The sticky, elastic mucus produced as a defensive secretion by the terrestrial slug *Arion subfuscus* is highly enriched in calcium, magnesium, zinc, iron, manganese and copper.[4,5]

It is likely that these proteins and metals act to cross-link the gel. Hydrogel mechanics are governed by polymer size, concentration and cross-linking.[1] The size and concentration of giant polymers control the mechanics of lubricating gels, but are not the primary factors influencing the properties of adhesive gels. The cohesive strength necessary for hydrogels to function as effective adhesives derives from cross-linking of the polymers.[1,2]

The presence of metals is exciting, as metals are capable of forming strong yet reversible cross-links in biomaterials.[6] Of particular importance is the fact that these bonds are often relatively insensitive to water; in contrast, common non-covalent interactions such as hydrogen bonds and electrostatic interactions are greatly weakened by the high dielectric constant of water.[4] Thus, metals have emerged as an important area of research in bioadhesion. The goal of this paper is to review the role of metals in biological adhesives with emphasis on the hydrogel adhesive of the terrestrial slug *A. subfuscus*. In addition, a mechanism is proposed that accounts for the presence of multiple cross-linking mechanisms and the marked strength of these gels.

2 TYPES OF METAL-BASED CROSS-LINKS IN BIOLOGICAL GLUES

2.1 Electrovalent and Coordinate Covalent Bonds

Metals can act both directly and indirectly to cross-link glues (Figure 1). Metals can serve as bridges to link polymers directly, using electrostatic or coordinate interactions, or they can catalyze oxidation reactions that create functional groups that readily form cross-links.[6] The simplest direct mechanism is electrostatic. Metal ions will be attracted to negative charges on polymers. In this way, a single divalent metal can bridge two negative charges on different polymers. Because of water's high dielectric constant, such interactions are inherently weaker in water.[7,8] Essentially, water molecules will orient around the charged groups and partially mask them. A stronger, more versatile interaction is through coordinate covalent bonds.[6] In these bonds, metals serve as electron pair acceptors, i.e. Lewis acids; they share electron pairs with donor ligands, i.e. Lewis bases. Lichtenegger et al. note that electrostatic interactions and coordinate covalent bonds can be viewed as part of a continuum, with bonds becoming more coordinate in nature as the ligands approach and lose associated solvent molecules.[6]

Many biological molecules contain potential metal-binding ligands. Biologically common donor ligands for hard Lewis acids such as Ca^{2+}, Mg^{2+}, Fe^{3+}, and Mn^{2+} include carboxyl groups, phosphate groups, sulfate groups and amino groups.[9] There is a wide

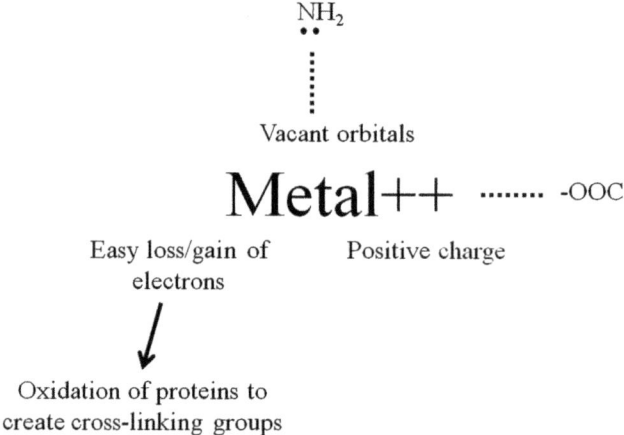

Figure 1   *Metal-based cross-linking mechanisms. Anionic groups such as carboxyls are attracted to the positive charge on metals and can thus interact electrostatically. Lewis bases such as amines can donate a pair of electrons to fill a vacant orbital on the metal and form a coordinate covalent bond. Finally, many metals can easily gain or lose electrons, thus catalyzing redox reactions. Metal-catalyzed oxidation of proteins commonly leads to carbonyl group formation, which provides a target for cross-linking via nucleophilic addition reactions. (Based on Lichtenegger et al.[6]).*

range of affinities, with metals such as Fe^{3+} typically forming stronger bonds than other metals, and ligands such as phosphate forming stronger bonds than other Lewis bases. Many biological glues are enriched in the rare amino acid 3, 4 dihydroxyphenylalanine, which can be oxidized into a dopa quinone. This has an unusually strong ability to coordinate hard acids such as Fe^{3+}.[10] A common ligand for borderline Lewis acids such as Zn^{2+} and Fe^{2+} is the imidazole group,[9] which is found on the side chain of the amino acid histidine. Sulphydryl groups are also common metal-binding groups, and these bind preferentially to soft Lewis acids, such as Cu^{+}.[11]

Coordinate covalent bonds have a number of useful properties. Even in aqueous solutions they approach the strength of common covalent bonds, but they are reversible.[6,12] The metals can dissociate and then reform the bond. This is significant because it raises the possibility that the material may heal after partial fracture. Additionally, if fracture proceeds slowly, metals may form new bonds with other ligands after separation from their initial ligand. This would increase the energy required to rupture the bond fully. Reversibility, however, also leads to lower stability of coordinate bonds in acid environments. This is because H^{+} is a hard Lewis acid that competes for donor ligands. Finally, a key feature of coordinate covalent bonds is that a single metal ion can be coordinated by multiple ligands. Thus, one protein can form multiple bonds to a single metal ion to hold it tightly, i.e. chelation. Alternatively, several different proteins can bind to the same metal, forming a cross-link.[6,10]

Metals have been implicated as part of cross-links in a number of biological materials. Calcium is involved in phosphoprotein crosslinking in Sabellarid tubeworm cement.[13-15] Proteins in the plaques, threads and outer coatings of mussel byssus are joined by metals such as calcium, copper and iron.[16-18] The hardened jaws of marine worms and different arthropods also often contain metals. In some cases these metals form coordinate covalent linkages,[6,19,20] while in others the metals are in mineral form.[6]

2.2 Metal Catalyzed Oxidation and Protein Cross-linking

Metals can also act catalytically to create functional groups that then go on to form cross-links. Metals are common co-factors of enzymes, and even when not bound to enzymes have the potential to drive reactions. The redox-active transition metals, especially iron and copper, are particularly important in this context. These metals can oxidize proteins, leading to substantial changes in their properties. Metal-catalyzed protein oxidation is a widespread phenomenon that typically leads to protein degradation and cellular pathology.[21,22] It can result from unwanted release of reactive oxygen species or from exposure to various external factors. This can result in damage that has been hypothesized to play a central role in aging.[21-24]

Interestingly, protein oxidation is not always damaging; it can also be harnessed to strengthen biomaterials. In the field of biological adhesives, most work on oxidation has centered on the rare and versatile amino acid 3,4-dihydroxyphenylalanine (dopa), which is highly enriched in a number of adhesives and biomaterials. Dopa is most famous as part of the proteins found in mussel byssus, where it serves as a cross-linking and adhesive agent.[8] In addition to their other properties, dopa and other catechols, as well as polyphenols, can be oxidized and then cross-linked through a number of mechanisms.[14] Dopa-metal binding with associated organic radicals has been detected in the byssus, thus providing strong evidence for oxidative reactions.[25] Catechols and low molecular weight polyphenols that are not necessarily part of proteins (ie. not peptidyl) are also involved in hardening a wide variety of materials including tunicate body walls, protozoan spores, soft coral skeletons, flatworm eggs, shark egg capsules, mussel byssus, snail periostracum, tubeworm tubes, earthworm eggs and arthropod cuticle.[26] In most cases, the copper-based enzyme catechol oxidase is responsible for oxidizing the catechol groups into quinones, which are then subject to a variety of cross-linking reactions.[26] These reactions have also been investigated in depth in arthropod cuticle.[27,28] The multicopper phenoloxidase laccase 2 is essential for hardening arthropod cuticle.[29] Once this enzyme oxidizes catechols, nucleophilic side chains on proteins attack the quinones to form conjugates. Polyphenols are also common in the attachment complexes of brown algae, where vanadium haloperoxidases seem to play a central role in creating larger cross-linked complexes.[30] Dopa and other phenolic compounds are particularly interesting because they can function by indirect and direct metal-based cross-links; they can be oxidized by metals to form cross-links, but without oxidation the dopa groups may chelate metals to serve in direct cross-links or interact with surfaces to generate adhesion.[7,14]

Dopa is not, however, the only amino acid that can be oxidized. There are eleven other common amino acids that are vulnerable to oxidation.[24] Of these, lysine, arginine and proline are common targets of oxidation. These yield carbonyl derivatives, specifically aminoadipic and glutamic semialdehydes.[24,31,32] These are modified amino acids that have flexible side chains terminating in carbonyl groups, which are often subject to cross-linking reactions (Figure 2).

The cross-linking of collagen and elastin is the best known example of this process. Oxidation of collagen and elastin by the copper-based enzyme lysyl oxidase is an essential step in cross-linking and strengthening animal connective tissue fibers.[33] Oxidation of lysine side chains leads to carbonyl groups tethered on flexible carbon chains. These can

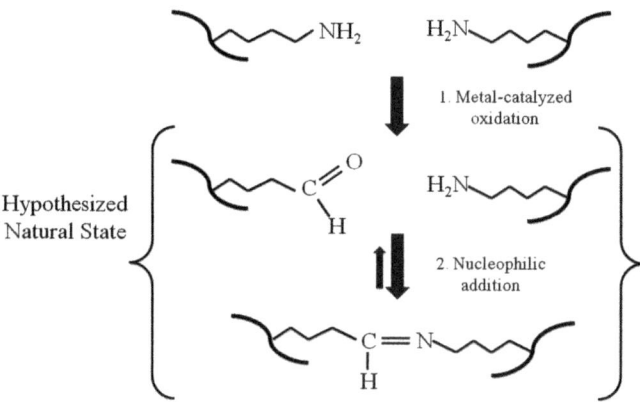

Figure 2   *Example mechanism for oxidative cross-linking of proteins proposed for A. subfuscus glue. Two lysine residues are shown on separate proteins (thicker lines). Metal catalyzed oxidation of some lysine residues leads to carbonyl formation. Other lysine residues can readily combine with these carbonyls by nucleophilic addition, leading to an imine bond, specifically a Schiff base. The reaction between the amine and the carbonyl is reversible, with its stability depending on conditions. Further modifications of the imine bond are also possible, creating permanent covalent bonds.*

react in a number of ways. They can react with other carbonyls to form aldol condensation products, or they can react with other lysine groups, leading to imine bond formation (Figure 2).[33] These are labile bonds that can be further modified in a variety of ways to create more stable cross-links.[33] Because iron and copper are common biominerals that are often part of oxidative enzymes, and they readily create cross-linking sites from common amino acids, it should not be surprising that this process has been harnessed to stiffen some biomaterials.

3 METAL-BASED CROSS-LINKS IN TERRESTRIAL SLUG GLUE

3.1 *Arion subfuscus* Glue Composition

The glue of the slug *Arion subfuscus* is a rapidly-setting hydrogel with substantial metal content.[4] The dorsal epithelium of the slug secretes this material in response to perceived threats. The material comes off the slug easily, but then sets rapidly into a highly elastic material that adheres strongly to the foreign surface. The adhesive secretion contains significant amounts of calcium (~40 mM), zinc (~3 mM), iron (~0.1 mM), manganese (~0.1 mM) and copper (~0.03 mM).[4,5] Magnesium is also common in the glue.[5] Calcium and zinc have been demonstrated to form direct cross-links in other biomaterials, so their high concentration is suggestive of such interactions. This is particularly noteworthy, as divalent and trivalent ions are known to have a large impact on polyanionic gels.[34] The presence of iron and copper also suggests that oxidation may be occurring. Iron, in fact, may serve in both roles, as it interacts strongly with multiple ligands as well as being redox active.

The polymer content of the glue consists of a mixture of giant polymers or polymer complexes and relatively smaller proteins in roughly similar quantities.[3] The larger polymers appear to be protein-polysaccharide complexes that are typical of most invertebrate mucus. What appear to make the glue special are the smaller proteins, especially specific proteins that are additionally present in large quantities in the glue as opposed to the non-adhesive secretion.[3]

The large polymers or polymer complexes appear to have relative molecular masses (M_r) larger than 1000×10^3 in gel filtration chromatography.[2,3] This component likely includes glycosaminoglycans. Sulfated glycosaminoglycans are common in molluscan mucus and are often called mucopolysaccharides.[35] They are likely a major component of *A. subfuscus* glue, based on this glue's high sulfate content.[5] They were also detected in a similar terrestrial slug.[36] Glycosaminoglycans are common components of secreted gels and visco-elastic extracellular tissues. They are often part of enormous complexes such as proteoglycans that are large enough to tangle at relatively low concentrations. This tangling is necessary for dilute materials to form loose gels.[1] The basic structure of invertebrate mucus typically consists of tangling networks of such polymers.

The remainder of the material in the glue consists of proteins with M_r ranging from roughly $14\text{-}200 \times 10^3$.[3] Two proteins, asmp-15 and asmp-61, have been identified as characteristic of the glue, as opposed to the non-adhesive trail mucus. The other proteins are equally present in the glue and the trail mucus, whereas the concentration of these two is markedly elevated in the glue.[3] Asmp-15 and a group of potentially related proteins near this mass (M_r 15×10^3) make up roughly half of the protein in the glue. These proteins bind to iron,[4] and have been shown to have the ability to stiffen several commercial gels composed of anionic polymers.[3] Asmp-61 has an M_r of 61×10^3, and its function has not been determined. In both the glue and non-adhesive trail mucus, there is a protein with an

M_r of 40 x 10^3. In the glue, this protein elutes from gel filtration as if it is part of a complex with M_r larger than 1000 x 10^3.[37] The apparent size of this protein from the non-adhesive mucus has not been studied; thus it is not clear if this protein becomes cross-linked as part of the adhesive process or is linked into complexes in all forms of mucus. It is notable that, unlike a number of other bioadhesives, the proteins in gastropod adhesive gels do not contain detectable amounts of the amino acid dopa.

3.2 Direct Cross-links in *A. subfuscus* Glue

Arion subfuscus glue appears to contain both direct and indirectly catalyzed metal-based cross-links. Direct interactions are easily demonstrated by the impact of metal chelators. Removing metals with EDTA causes a dramatic increase in solubility and a similar decrease in stiffness.[5,37] Metal removal destroys the integrity of the glue, but when metals are present, the glue is relatively insoluble and tough. Further work implicates calcium in these cross-links. Buffers that preferentially bind to hard Lewis acids like calcium weaken the glue, while buffers that bind with zinc do not.[5] Similarly, chelators with a high affinity for transition metals but low affinity for calcium did not affect the glue's solubility or mechanics.[4,5] Taken together with the fact that calcium is the most common metal ion in the glue, and that its total charge roughly balances that contributed by sulphate it is clear that calcium plays a central role in the glue structure.[5] Many other commercial gels similarly depend on ions like calcium, such as pectin, carrageenan and alginate.[38] These are charged, polysaccharide gels. As with these gels, calcium will be electrostatically attracted to the charged groups on the polysaccharides. Nevertheless, it is worth noting that none of these commercial gels are effective adhesives.[1]

It is unlikely that electrostatic interactions between calcium and glycosaminoglycans alone explain the mechanics of the glue. Disrupting electrostatic interactions with high salt concentrations had no impact on the solubility of the glue.[37] Most molluscan mucus secretions contain large polyanions, and there must be counterions to balance that charge. Calcium has been described in many invertebrate secretions, and there is no difference in the calcium or magnesium content between the adhesive and non-adhesive mucus of the limpet *Lottia limatula*.[39] Thus, calcium and glycosaminoglycans may form an essential backbone to many secretions, and disrupting either would disrupt the glue, but they may not provide the additional strength needed to convert from a lubricating to an adhesive function.

It is likely that the added strength comes from stronger coordinate covalent linkages, and/or cross-links derived from metal-catalyzed oxidation. Such linkages likely involve asmp-15 and asmp-61, the proteins that characterize the adhesive form of the gel. Asmp-15 binds to iron.[4] Thus, it likely also can bind to calcium, as both Ca^{2+} and Fe^{3+} share the same preferred ligands. If these are coordinate linkages, they would likely have a stronger metal-binding ability than glycosaminoglycans. Thus, adding asmp-15 to the glue and forming coordinate linkages could create a stiffer network. Also, if the new cross-links involve iron, that might strengthen the glue more than calcium. Metal-binding ligands strongly prefer binding iron to binding calcium, following the stability series known as the Irving-Williams series.[11] Based on this series, manganese, which is also present in the glue, would also form stronger interactions than calcium or magnesium. Holten-Andersen et al. emphasize that even a small amount of iron can have a substantial effect because of its high affinity for some ligands,[40] and Hwang et al. propose that both calcium and iron contribute to the strength of the mussel byssus via direct cross-links, with calcium being much more prevalent but forming weaker cross-links.[41] On the other hand, the inability of the transition metal chelator deferoxamine to dissolve mature slug glue suggests that iron, zinc and manganese do not contribute to direct cross-links. Nevertheless, that could be explained if the proteins had a higher affinity than the chelator, so it was unable to disrupt the interactions.[4] In fact, EDTA was not able to remove iron from the proteins in the glue.[5] This strongly suggests high affinity coordinate linkages involving iron.

The relatively easier removal of zinc, combined with the fact that its removal did not weaken the gel suggests that zinc does not strengthen the glue directly. This is surprising given its importance in other biomaterials and its high concentration in the gel. It is likely zinc plays some other, currently undetermined role. Both zinc and manganese could play a modulatory role, as they can compete with iron for substrates and thus impede their oxidation.[24,42]

Overall, it seems likely that calcium, manganese and iron all contribute directly to the integrity of the glue, with calcium playing an essential role, though perhaps one that occurs in many mucous secretions, whether they are adhesive or not. The increased strength of the adhesive may come by direct cross-links involving proteins with higher affinity metal-ligand interactions, as well as oxidatively derived cross-links. It is unclear whether these involve only the proteins, or the proteins and the polysaccharides.

3.3 Oxidative Cross-links in *A. subfuscus* Glue

Recent work has provided evidence for oxidatively derived cross-links in slug glue. Several of the common proteins in the glue are heavily oxidized.[43] In fact, one protein was as heavily oxidized as a positive control that had been incubated with ferrous iron and hydrogen peroxide. The oxidized proteins appear to form imine bonds (Fig. 2), and these bonds contribute to the mechanical strength of the glue. When treated with a strong nucleophile to disrupt these bonds, the glue was less stiff,[5] and the oxidized proteins were more readily dissolved.[43] When treated with sodium borohydride to reduce and thus stabilize imine bonds, the glue became stiffer and the oxidized proteins were no longer soluble.[5,43] All of the proteins other than asmp-15 were affected, especially those with M_r between 57-200 x 10^3 suggesting that they form an imine cross-linked network.[43] Since imine bonds are labile, this network would be soluble in SDS, but once reduced by borohydride, would form an insoluble, permanent network. The imine bonds detected in these experiments may actually represent only a part of the oxidatively derived bonds, as such bonds are often modified further and become permanent.[33]

There may be great value in the labile nature of imine bonds. Reversible bonds are useful for their ability to reform when the stress is removed, allowing the material to heal.[10,44] There is also significant potential to control the equilibrium between the imine bond and the precursor carbonyls and amines, thus controlling performance.[45]

While most of the proteins appear to be part of an imine-bonded network, asmp-15 may have a different function. These iron-binding proteins are present in larger quantities than any of the other proteins, and they appear to play a central role in the glue. They may contribute by coordinate covalent cross-linking, but they may also play a catalytic role. They have the ability to non-specifically stiffen anionic polymer gels, and this activity depends on the presence of metals.[3,4] This may be due to a catalytic event such as oxidation, or it may just reflect the formation of direct cross-links. Evidence for a catalytic event comes from the finding that oxidation appears to occur after secretion of the glue.[43] Furthermore, the effect of the transition metal chelator deferoxamine is much greater during the process of glue setting. It does not affect the solubility or stiffness of mature glues,[4,5] but when it is present as the glue is secreted, the polymers disperse immediately into liquid rather than setting into a gel.[4] Collecting glue into deferoxamine also leads to a decrease in the tendency of asmp-40 to form large complexes.[37]

While this evidence suggests a catalytic event, it does not rule out direct cross-links. Glue collected into chelators still contains oxidized proteins, so perhaps deferoxamine acts on the glue in a different way.[43] It may be able to bind to metals before they can form strong coordinate bonds with their natural ligands, but not once the bonds form. The large quantity of asmp-15 also suggests a more direct mechanical role than an enzymatic role. One would not normally expect an enzyme to make up a large fraction of a secretion. The large quantity may be necessary, though, to keep sufficient iron available to drive the reaction. It may also be necessary to compensate for the fact that protein mobility will be low in such a densely structured, cross-linked gel, thus potentially limiting the rate at which an enzyme can act.[46]

3.4 Proposed Model for *A. subfuscus* Glue

Sulfated polysaccharides and calcium likely form the backbone of the gel, and their electrical interaction would certainly play a key role in determining the overall structure. Nevertheless, as discussed previously, this alone may not be sufficient to create a strong glue. Instead, it probably forms a scaffolding for the gel, or a separate network. Perhaps electrostatic interactions bring this network together, then organize the proteins. The proteins may then form cross-links with each other, or with polysaccharides by additional mechanisms.

The way the glue components are secreted and mixed is likely to be important. As with baking, it is not enough to know the ingredients of a cake; the manner in which they are assembled matters. A coacervation mechanism similar to the one proposed for tubeworm cement by Stewart et al.[13] may help create the overall structure of the glue. This mechanism depends on charge balancing between polyanions and polycations. Sulfated glycosaminoglycans are strong polyanions. Biochemical analyses have suggested that most of the proteins have acidic isoelectric points and would also be anionic.[2] Nevertheless, histochemical analyses have found basic proteins in the adhesive glands of slugs.[47,48] In addition, all the gel-stiffening, adhesive-specific proteins that have been found in gastropods, such as asmp-15, have 8-9% lysine content, which is significantly higher than any of the other proteins.[39,43,49] Thus, they may have polycationic regions that may interact with the sulfated carbohydrates, or with the more acidic proteins. Divalent ions such as calcium and magnesium are available to balance excess negative charge. Such shielding is necessary in secretory granules.[50] Electrostatic interactions could bring together the polymers in a complex network consisting of fiber-rich regions and dilute, aqueous intervening regions. Stewart et al. emphasize that this kind of electrostatic interaction is likely, but that the lack of complete mixing in sandcastle worm and caddisfly larva cements suggests that a complex coacervation mechanism leading to multiple levels of macromolecular structure as originally proposed may not be acting.[51]

The main components appear to be in separate glands, which raises the possibility of mixing and structural rearrangement after secretion. Three glands have been identified in the dorsal epithelium of the related slug *Ariolimax columbianus* and another pulmonate gastropod, *Helix aspersa*. One gland, the channel cell or protein gland, secretes mostly fluid but possibly also protein. Another gland, the mucus gland, secretes acid polysaccharides. The third contains calcium with proteins.[47,52,53] In *A. columbianus*, the channel cell is reported to be eosinophilic, containing basic proteins, while the calcium gland in *H. aspersa* appears to contain basic protein.[47] In either case, the basic protein is separate from the main mucopolysaccharide glands. The polysaccharides themselves are likely associated with divalent ions, even though that did not show up on the histology. It is worth noting that most histological methods only detect mineral calcium, or calcium ligands, rather than total calcium.[54] The secretions themselves are packaged in granules that flow easily until ruptured by some stimulus.[55] This would allow mixing. When the basic proteins can mix with polyanionic glycosaminoglycans, they may displace associated ions. This would likely be a favored reaction due to the entropy increase associated with the polycationic region of a single protein displacing multiple ions from sulfate on the polysaccharide.[46] In addition, having multiple ligands on one polymer greatly decreases the probability of dissociation relative to ions, making it more stable.[46] Thus, there is the potential for separate components mixing after their secretory packets rupture, and rapidly reorganizing to form a stronger network.

After bringing all the components together, further cross-linking can occur. As mentioned previously, coordinate covalent bonds and oxidatively-derived imine bonds seem to play a role. Glue-specific proteins such as asmp-15 may form a metal-coordinated network, and/or be involved in the formation of an imine-bonded network. This process of electrostatically driven coacervation followed by further cross-linking is similar to what occurs in sandcastle tubeworm cement.[46] In sandcastle worms, though, dopa plays a key role, while dopa has not been detected in gastropod glue.[39,43,49] Also, sandcastle cement is strongly affected by the pH shift between the secretory granules and seawater,[46,51] whereas slug glue is secreted into air. Thus, the general pattern may be similar, but the specifics are different.

Finally, it is important to note that the proteins may be brought together and then cross-link with the polysaccharides, or they may interpenetrate the polysaccharide network, but not physically link to it. This turns out to be an important distinction; are the proteins and polysaccharides all linked together into a single network involving multiple cross-linking mechanisms, or do they form separate networks? This has important implications for mechanical performance because two interpenetrating networks can work synergistically.

3.5 Multiple Cross-linking Mechanisms and Double Network Gels

The presence of multiple cross-linking mechanisms involving different types of polymers raises the possibility that slug adhesive gels are double network gels as described by Haque et al.[56] In these gels, having two networks of polymers within the same material can result in several orders of magnitude greater toughness than the individual networks on their own. A key feature is that the two networks differ markedly in features such as polymer size, cross-linking density and stiffness.[56] For example, great increases in toughness occur when a stiffer, highly cross-linked network is embedded within a softer, less cross-linked network of larger polymers.[57-59] Rupture between cross-links in the more rigid network occurs first, but the second network remains intact. The second network redistributes stress concentrations, making it difficult for the crack to propagate.[58] Also, the second network must still be deformed until sufficient strain energy accumulates to break it. In the process, the extra deformation forces further damage to the first network.[57] The original network may break into clumps throughout a large zone, whose further deformation depends on the "hidden length" of the second ductile network (Figure 3).[56,57] Thus, instead of a crack traveling easily through a plane, the damage and deformation occurs through a larger volume of material.[58] This requires a great deal more energy to fracture the material.

One way to think of this mechanism is to consider different single network gels, such as gelatin and mucus. Gelatin alone forms stiff, highly cross-linked gels, but it is brittle; fractures propagate easily through it because there is no effective dissipative mechanism. Lubricating forms of mucus, with their giant, tangled polymers, form weak, easily

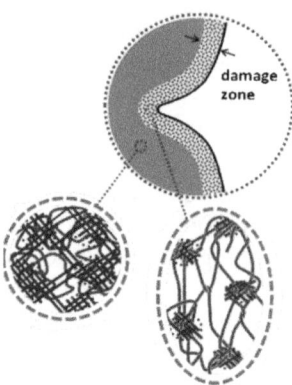

Figure 3   *Fracture of a double network gel. In the undamaged zone, a tightly cross-linked network is set within a ductile network of much larger, loosely linked polymers. As fracture occurs, cross-links within the former network break, but a great deal of further deformation is required to disrupt the loose network of larger polymers. As that occurs, the looser network redistributes stress and the large strain leads to further fracture within the more tightly cross-linked network. Thus, the damage zone extends through a large volume in the vicinity of the crack. (Reproduced by permission of the Royal Society of Chemistry from ref. 59).*

deformable gels, which dissipate energy well but have little stiffness. Combining two such gels can benefit from their complementary strengths.

The conditions necessary for the toughness enhancement of double network gels seem well-matched to the structure of molluscan adhesive gels, and many loose connective tissues in general. The presence of giant, loosely cross-linked polymers is almost universal in mucus and similar gels. Interpenetrating that network of giant polymers with a more highly cross-linked protein network would likely lead to dramatic increases in toughness. One of the main findings to come out of the study of adhesive gels is that they are similar to mucus secretions, but additionally contain large quantities of relatively smaller proteins that play a central role in the mechanical change.[3,39,49] This fits with the double network model. The additional proteins do not merely add strength proportional to their extra cross-links; they may fundamentally change the gel by adding a second, interpenetrating network with different properties. If molluscan adhesive gels do achieve markedly greater toughness through double networks, it would represent the first description of this mechanism in nature. All previous work on double network gels has involved the study of synthetic gels.

This double network mechanism could explain the exceptional performance of the adhesive gels of terrestrial slugs. The giant polymers typical of mucus would form the ductile network, and then the addition of a more heavily cross-linked protein network would toughen this far in excess of what would be expected of the two networks acting separately. A similar mechanism may operate in brown algal holdfasts, which are described as a mixture of polysaccharides, specifically calcium cross-linked alginates, and cross-linked lower molecular weight phenolics.[30]

In slug glue, the fact that the bonds in the heavily cross-linked protein network are reversible increases the value of this system. Current synthetic double network gels involve irreversible bond breakage, limiting the material's fatigue resistance. Haque et al. suggest that this could be avoided if suitable reversible bonds could be found.[56] That may be what we see in the slug's adhesive gel, which depends on metal coordination and imine bonds, both of which are strong, covalent bonds that are also reversible.

Despite the similarities, there are notable differences between gastropod adhesive gels and the synthetic double network gels. In the synthetic gels, a common criterion for effective gel toughening is that the ductile network should consist of neutral polymers,[57] whereas the proposed ductile network in slug glue likely depends on polyanionic glycosaminoglycans. Synthetic double network gels are also typically more concentrated, with about 10% polymer content.[58] There are other criteria, such as inhomogeneity in the networks,[58] and further work will be necessary to see how well gastropod adhesive gels match these criteria. It is possible that they form a newer class of double network gels, with slightly different properties and mechanics. Their study may then suggest new approaches in the design of synthetic double network gels.

4 CONCLUSION

A. subfuscus glue is intriguing due to the combination of different polymers with multiple types of cross-links. Metals are essential to the performance of the glue, creating strong, covalent but reversible bonds. Hard Lewis acids such as calcium and possibly iron form direct cross-links, while redox active metals likely drive the oxidation of several of the proteins so that they can react with amines to form imine bonds and possibly other cross-links that can form from protein carbonyls.

A key question is which polymers participate in each type of cross-link. Glycosaminoglycans likely form large complexes and interact via calcium. They may also interact with polycationic regions of proteins and proteins that may have calcium-binding regions. The imine bond network appears to form primarily among proteins ranging in M_r from roughly 57-200 x 10^3. The primary adhesive protein, asmp-15, binds to iron, and this may form strong coordinate bonds, but may be involved in metal-catalyzed oxidation. The combination of large polymers in a loosely cross-linked network, and smaller, more heavily cross-linked polymers may function as a double network to achieve much greater toughness than a simple network based on a single cross-linking mechanism.

References

1. A. M. Smith, *Integr. Comp. Biol.*, 2002, **42**, 1164.
2. A. M. Smith in *Biological Adhesives,* ed. A. M. Smith and J. A. Callow, Springer, Berlin, 2006, p 167.
3. J. M. Pawlicki, L. B. Pease, C. M. Pierce, T. P. Startz, Y. Zhang and A. M. Smith, *J. Exp. Biol.,* 2004, **207**, 1127.
4. S. W. Werneke, C. Swann, L. A. Farquharson, K. S. Hamilton and A. M. Smith, *J. Exp. Biol.,* 2007, **210**, 2137.
5. M. Braun, M. Menges, F. Opoku and A. M. Smith, *J. Exp. Biol.,* 2013, **216**, 1475.
6. H. C. Lichtenegger, H. Birkedal and J. H. Waite, *Met. Ions Life Sci.*, 2008, **4**, 295.
7. J. H. Waite, *Integr. Comp. Biol.*, 2002, **42**, 1172.
8. J. H. Waite, N. Holten-Andersen, S. A. Jewhurst and C. J. Sun, *J. Adhes.*, 2005, **81**, 1.
9. T. L. Ho, *Chem. Rev.*, 1975, **75**, 1.

10. M. J. Harrington, A. Masic, N. Holten-Andersen, J. H. Waite and P. Fratzl, *Science*, 2010, **328**, 216.

11. S. L. Lippard and J. M. Berg, *Principles of Bioinorganic Chemistry*, University Science Books, Mill Valley, 1994.

12. N. Holten-Andersen, M. J. Harrington, H. Birkedal, B. P. Lee, P. B. Messersmith, K. Y. Lee and J. H. Waite, *Proc. Natl. Acad. Sci. U.S.A.*, 2011, **108**, 2651.

13. R. J. Stewart, J. C. Weaver, D. E. Morse and J. H. Waite, *J. Exp. Biol.*, 2004, **207**, 4727.

14. J. Sagert, C. Sun and J. H. Waite in *Biological Adhesives,* ed. A. M. Smith and J. A. Callow, Springer, Berlin, 2006, p 125.

15. C. J. Sun, G. E. Fantner, J. Adams, P. K. Hansma and J. H. Waite, *J. Exp. Biol.*, 2007, **210**, 1481.

16. E. Vaccaro and J. H. Waite, *Biomacromolecules*, 2001, **2**, 906.

17. H. Zhao and J. H. Waite, *Biochemistry*, 2006, **45**, 14223.

18. M. J. Harrington and J. H. Waite, *J. Exp. Biol.*, 2007, **210**, 4307.

19. H. C. Lichtenegger, T. Schoberl, J. T. Ruokolainen, J. O. Cross, S. M. Heald, H. Birkedal, J. H. Waite and G. H. Stucky, *Proc. Natl. Acad. Sci. U.S.A.*, 2003, **100**, 9144.

20. C. C. Broomell, M. A. Mattoni, F. W. Zok and J. H. Waite, *J. Exp. Biol.*, 2006, **209**, 3219.

21. E. R. Stadtman and C. N. Oliver, *J. Biol. Chem.*, 1991, **266**, 2005.

22. C. L. Hawkins, P. E. Morgan and M. J. Davies, *Free Radical Biol. Med.*, 2009, **46**, 965.

23. E. R. Stadtman, *Science*, 1992, **257**, 1220.

24. B. S. Berlett and E. R. Stadtman, *J. Biol. Chem.*, 1997, **272**, 20313.

25. M. J. Sever, J. T. Weisser, J. Monahan, S. Srinivasan and J. J. Wilker, *Angew. Chem. Int. Ed.*, 2004, **43**, 448.

26. J. H. Waite, *Comp. Biochem. Physiol. B.*, 1990, **97**, 19.

27. T. L. Hopkins and K. J. Kramer, *Annu. Rev. Entomol.*, 1992, **37**, 273.

28. K. J. Kramer, M. R. Kanost, T. L. Hopkins, H. Jiang, Y. C. Zhu, R. Xu, J. L. Kerwin and F. Turecek, *Tetrahedron*, 2001, **57**, 385.

29. Y. Arakane, S. Muthukrishnan, R. W. Beeman, M. R. Kanos and K. J. Kramer, *Proc. Natl. Acad. Sci. U.S.A.*, 2005, **102**, 11337.

30. P. Potin and C. Leblanc in *Biological Adhesives* ed. A. M. Smith and J. A. Callow, Springer, Berlin, 2006, p 105.

31. A. Amici, R. L. Levine, L. Tsai and E. R. Stadtman, *J. Biol. Chem.*, 1989, **264**, 3341.

32. J. R. Requena, C.-C. Chao, R. L. Levine and E. R. Stadtman, *Proc. Natl. Acad. Sci. U.S.A.*, 2001, **98**, 69.

33. M. L. Tanzer, *Science*, 1973, **180**, 561.

34. T. Tanaka, *Sci. Am.*, 1981, **244**, 124.

35. M. W. Denny in *The Mollusca.* Vol. I, ed. K. Wilbur, K. Simkiss and P. W. Hochachka, Academic Press, New York, 1983, p 431.

36. J. M. Cottrell, I. F. Henderson, J. A. Pickett and D. J. Wright, *Comp. Biochem. Physiol. B*, 1993, **104**, 455.

37. A. M. Smith, T. M. Robinson, M. D. Salt, K. S. Hamilton, B. E. Silvia and R. Blasiak, *Comp. Biochem. Physiol. B.*, 2009, **152**, 110.

38. G. O. Phillips and P. A. Williams, *Handbook of hydrocolloids*, Woodhead Publishing Limited, Cambridge, 2000.

39. A. M. Smith, T. J. Quick and R. L. St. Peter, *Biol. Bull.*, 1999, **196**, 34.

40. N. Holten-Andersen, T. E. Mates, M. S. Toprak, G. D. Stucky, F. W. Zok and J. H. Waite, *Langmuir*, 2009, **25**, 3323.
41. D. S. Hwang, H. Zeng, A. Masic, M. J. Harrington, J. N. Israelachvili and J. H. Waite, *J. Biol. Chem.*, 2010, **285**, 25850.
42. M. Chevion, *Free Radical Biol. Med.*, 1988, **5**, 27.
43. A. Bradshaw, M. Salt, A. Bell, M. Zeitler, N. Litra and A. M. Smith, *J. Exp. Biol.*, 2011, **214**, 1699.
44. H. Zeng, D. S. Hwang, J. N. Israelachvili and J. H. Waite, *Proc. Natl. Acad. Sci. U.S.A.*, 2010, **107**, 12850.
45. M. E. Belowich and J. F. Stoddart, *Chem. Soc. Rev.*, 2012, **41**, 2003.
46. R. J. Stewart, T. C. Ransom and V. Hlady, *J. Polym. Sci. B Polym. Phys.*, 2011, **49**, 757.
47. M. Campion, *Q. J. Microsc. Sci.*, 1961, **102**, 195.
48. A. W. Martin and I. Deyrup-Olsen, *J. Exp. Biol.*, 1986, **121**, 301.
49. A. M. Smith and M. C. Morin, *Biol. Bull.*, 2002, **203**, 338.
50. C. S. Wang and R. J. Stewart, *J. Exp. Biol.*, 2012, **215**, 351.
51. R. J. Stewart, C. S. Wang and H. Shao, *Adv. Colloid Interface Sci.*, 2011, **167**, 85.
52. D. L. Luchtel, A. W. Martin and I. Deyrup-Olsen, *Cell Tissue Res.*, 1984, **235**, 143.
53. A. M. Smith in *Biological Adhesive Systems: From Nature to Technical and Medical Application,* ed. J. v. Byern and I. Grunwald, Springer, Berlin, 2010, p 41.
54. G. L. Humason, *Animal Tissue Techniques*, W. H. Freeman and Company, San Francisco, 1979.
55. I. Deyrup-Olsen, D. L. Luchtel and A. W. Martin, *Am. J. Physiol.*, 1983, **245**, R448.
56. M. A. Haque, T. Kurokawa and J. P. Gong, *Polymer*, 2012, **53**, 1805.
57. H. R. Brown, *Macromolecules*, 2007, **40**, 3815.
58. R. E. Webber, C. Creton, H. R. Brown and J. P. Gong, *Macromolecules*, 2007, **40**, 2919.
59. J. P. Gong, *Soft Matter*, 2010, **6**, 2583-2590.

THE MINERALIZED BYSSUS OF *ANOMIA SIMPLEX*: A CALCIFIED ATTACHMENT SYSTEM

H. Birkedal,* S. Frølich, H. Leemreize, R. Stallbohm, Y.H. Tseng

iNANO and Department of Chemistry, Aarhus University, Denmark.
*hbirkedal@chem.au.dk

1 INTRODUCTION

Bivalve molluscs adopt a wide range of strategies for attachment to substrates. Prime amongst these is the byssus best known from the widespread mussels (*Mytilidae*), but found in a large number of adult bivalves and an even larger number of bivalves during development.[1] In the mussels, a large number of individual byssal threads form the byssus apparatus.[2-5] The threads are organic in nature and predominantly made of proteins. Their graded mechanical design[6] and the presence of DOPA in both byssus thread and adhesive pad[2-5] have inspired several researchers to develop synthetic functional materials inspired by mussel byssi.[2,7] In contrast to the protein based 'soft' byssus of the mussels, the members of the Anomiidae family display calcified, i.e. biomineralized by $CaCO_3$, byssi that are much less understood. In the present chapter we review the literature on the structure of these byssi and present new data on the magnesium distribution in the inorganic part of these peculiar attachment systems. The chapter is organized as follows: first we provide an overview of the Anomiidae followed by a more detailed discussion of the best studied members of the family namely *Anomia* sp. This is followed by a description of experiments performed to determine the magnesium distribution in the byssi after which the chapter finishes with a discussion of open questions.

2 CALCIFIED BYSSI: OVERVIEW OF THE ANOMIIDAE

The Anomiidae are tightly attached to their substrate through a byssus attachment system. An overview of the family has been given by Yonge and will therefore be kept brief.[8] The family consists of four genera: *Anomia*, *Pododesmus*, *Heteranomia* and *Enigmonia*. In all members of the family, except *Enigmonia*, the byssus is calcified. The byssus extends through the byssus notch that is placed in the right valve (Figure 2A and 2B). In contrast to most other bivalves, the animals are highly asymmetric: the tight byssal attachment results in the right valve being placed towards the substrate to such a degree that the morphological right and left valves *in situ* topographically occur as bottom and top shells, respectively. This can be seen in Figure 1 and 2 that show *Anomia simplex* animals *in situ* attached to substrates and an animal pulled off the substrate, respectively. The animals

Figure 1   Anomia simplex *specimens attached to a shell (A) and pebbles (B). Note how the animals adapt their shape to the substrate and how they snuggly fit onto the substrate.*

follow the local substrate topography so that even the top shell displays the shape of the underlying material as exemplified by the animals in Figure 1. The very large byssus, Figure 2, is connected to the top shell via a large retractor muscle. The bottom and top valves are connected by an adductor muscle, the scar of which can be seen in the photograph of the bottom shell in Figure 2B, see also the sketch in Figure 3C. This adductor is significantly smaller than the byssus in all species[8] suggesting that the byssus also plays a role in shell closure and not only in attachment.

The different members of the Anomiidae differ in terms of their typical size, the number of muscle scars in the left valve and several other anatomical details[8]. *Pododesmus* sp. are large animals with shells attaining sizes larger than 10 cm while *Anomia* sp. are smaller, typically below 5 cm. All members with the exception of *Enigmonia* sp. live from tidal to subtidal depths. Attachment in both *Pododesmus* and *Anomia* is permanent in the adult, but may be reformed in young individuals where attachment is still temporary.[1,9]

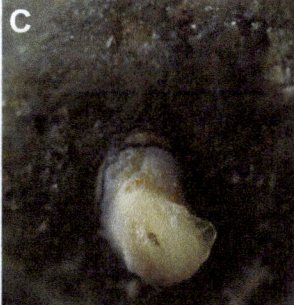

Figure 2   Anomia simplex *shell and byssus. Panel A shows an animal pulled off the substrate revealing the large byssal notch in the right valve, which is the one facing the substrate in vivo. The scale bar is 1 cm. B shows the interior surface of the right valve showing the byssal notch marked with b and the adductor muscle scar marked with s. Note the different shell structure around the byssal notch and adductor scar. C byssus from the animal shown in A still attached to the pebble the animal was originally attached to. The semi-transparent mass visible at the top of the byssus column is sheets of organic material that interleaves into the musculature of the animal.*

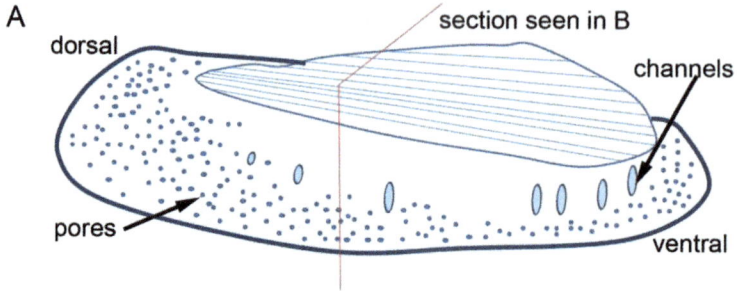

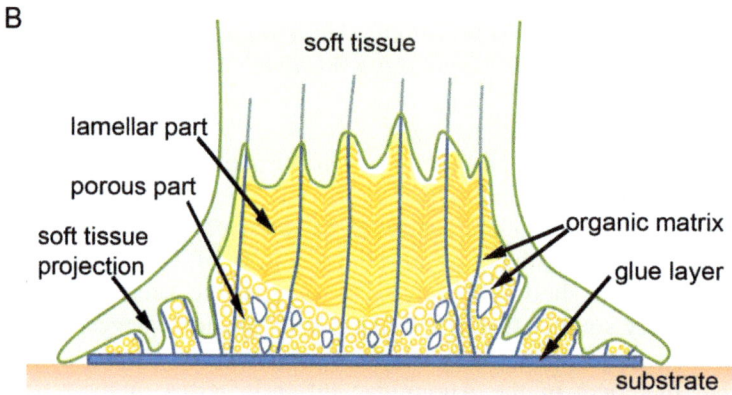

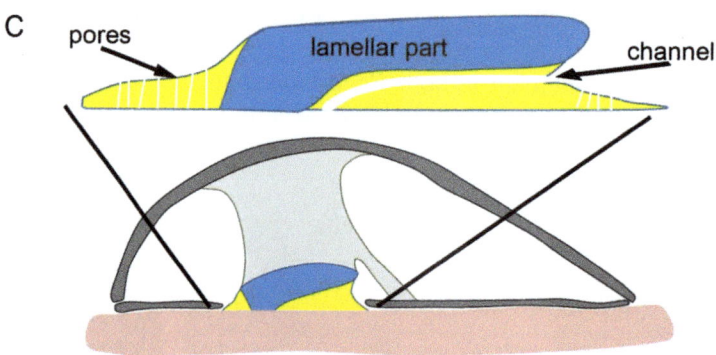

Figure 3   *Sketch of byssus design based on the works of Yamaguchi[9] and of Eltzholtz and Birkedal.[11] (A) shows the mineralized part of the byssus apparatus. The view shown in (B) represents a cross section perpendicular to the dorsal ventral direction. (C) shows a sketch of a cross section parallel to the dorsal ventral direction. Note that the number of lamellae and the size of structural units are not to scale. The bottom part of sketch C shows the placement in the entire animal. For a full description see text.*

Enigmonia is unique amongst the Anomiidae in that the byssus is not mineralized and the animal remains mobile in adulthood.[8,10] It lives on the stems, branches and leaves of mangroves and appears to use firm attachment as a means to avoid desiccation when the tide retracts and exposes the animals to air.

3   THE BYSSUS OF *ANOMIA*

The byssus structure of *Anomia* has been investigated by several authors.[8,9,11-13] It consists of a lower part that is heavily calcified, over 90 wt% $CaCO_3$[14], and an upper part that forms the interface with the musculature. Figure 3 shows a sketch of the byssus in an adult individual with Figure 3A displaying the mineralized portion only. The mineralized part of the byssus construct consists of a lamellar and a porous part.[9,11,12]

The lamellar portion forms the interface towards the musculature (soft tissue) via a rippled interface consisting of hills and valleys that are continuous in the dorsal-ventral direction, leading to varying heights perpendicular to this direction (Figure 3B). This interface design presumably acts to increase the surface area of interaction between soft tissue and the mineralized byssus. The lamellar part extends from the interface between byssus and musculature towards the substrate. Along the way it gradually becomes smaller in area. It consists of sheets of organic matrix that interleave from the mineralized lamellar part into the soft tissue of the animal (Figures 3B and 3C). These sheets of organic matrix are decorated with a herringbone arrangement of needle shaped crystals and/or crystal assemblies[11] of aragonite[9] as shown in Figure 4. The herringbone crystals change direction when moving up through the byssus and even branch into two sets of herringbone close together[11]. Between the herringbone decorated organic sheets, one finds additional crystals described as calcitic spherulitic needles in *Anomia chinensis* by Yamaguchi[9], but seen as assemblies of rods by Eltzholtz *et al.* in *Anomia simplex*.[11,13] The porous part consists of irregular spheroidal assemblies of rod/needled crystals interspersed with organic matrix.[11,13] The average size of the spheroidal assemblies changes with distance from the substrate interface from ~5 μm towards the interface to ~35 μm closer to the lamellar part, Figure 3B. At the interface between lamellar and porous parts, the spheroidal assemblies

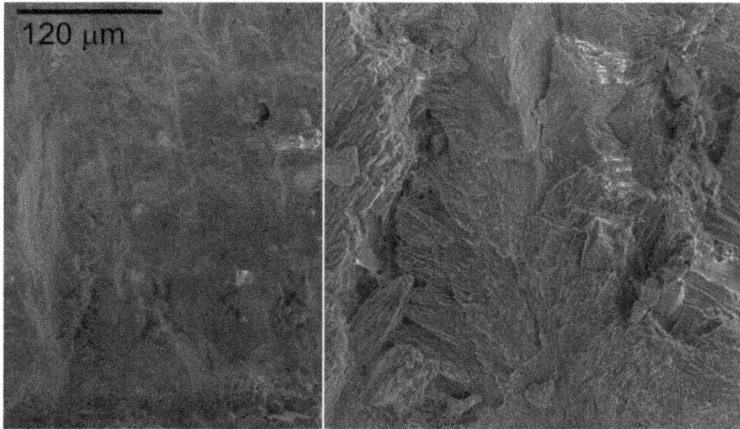

Figure 4   *SEM image of a fractured byssus showing part of the herringbone pattern of the lamellar part of the byssus adapted from Eltzholtz and Birkedal.[11] The image on the left is taken so that the bottom part is at the border of the porous and lamellar part of the byssus. The right image is an higher magnification (4 times higher) subset of the figure on the left.*

are again small[11] creating a distinct change in optical aspect when viewed under an optical microscope.

The porous part is penetrated by vertical pores, Figure 3, that extend from the surface of the byssus to the substrate.[9,11,12] The pores are on the order of 10-30 μm in diameter. They are lined by light brown organic matrix that in turn is formed by sheets that appear to be glued together by another structure.[11] The light brown colour is reminiscent of protein tanning and we speculate that such a tanning procedure may be present and aid in ramifying the organic network. Larger channels extend into the centre of the byssus from the ventral side moving through the porous layer parallel to the substrate finally turning downwards towards the substrate, Figure 3C. In larger byssi, there can be several layers of channels.

Inside the living animal, the byssus is covered by byssal gland tissue that forms projections into the pores of the byssus[9] as indicated by the light grey area marked 'soft tissue' in Figure 3B. Yamaguchi proposed that the pores may allow glue to reach the byssus substrate interface and to generate a larger contact area between gland tissue and byssus.[9] In the adult animal the foot is significantly reduced in size and no longer supports locomotion, but is rather thought to serve to clean the byssus/shell surfaces.[8]

Interestingly, when the musculature pulls the top shell towards the byssus, it will subject the biomineralized byssus to tension and the byssus is thus one of the few biomineralized structures operating predominantly under this type of mechanical load. The complex hierarchical arrangement of crystals and organic phases in the byssus most likely results in a controlled set of mechanical properties. Interfacing soft musculature to stiff rocks and shells requires specialized structures to avoid building up excessive interfacial strain that may result in catastrophic failure at the stiff/soft interface.[6] The retractor mussels pull on the organic sheets that are dovetailed into the mineralized byssus via the rippled lamellae/soft tissue interface. We speculate that the lamellar structure results in a locally varying stiffness of the byssus that in turn will mitigate stress build up. Detailed investigations of mechanical properties are underway in our laboratory to address this hypothesis.

3.1 Byssus Formation

The size of the byssus scales with the size of the animal. Yamaguchi studied byssus formation in *Anomia chinensis*.[9] Byssus development takes place in five distinct stages as illustrated in Figure 5. The settling animal first deposits a small organic disk harbouring lamellae in stage I. Vestiges of this initial attachment site remains in the adult animal (barring possible movement during the further development). In stage II the byssus starts to calcify with aragonite forming at the lamellae with inter-lamellar spherulites of calcite also forming. Thereafter (stage III) the mineralized lamellae increase in thickness and increase in height ventrally. At stage IV the porous layer starts forming on the ventral side of the byssus with the channels (Figure 3C) and pores appearing as well. In the final, adult byssus (stage V), the structure is further ramified by the porous layer that now surrounds the entire base of the lamellar part. The animals can return to the crawling state, abandoning the forming byssus, anywhere up to and including stage III. Once the porous layer starts forming, recementation is not observed and animals perish upon loss of the byssus. During stage IV, the byssus can however still be moved laterally while the animal remains in place. During adulthood, the byssus is often seen to rotate by turning the lamellae. This behaviour was proposed by Yamaguchi to be an adaptation to the shape of the substrate (see the detailed adaptation to substrate morphology in Figure 1).

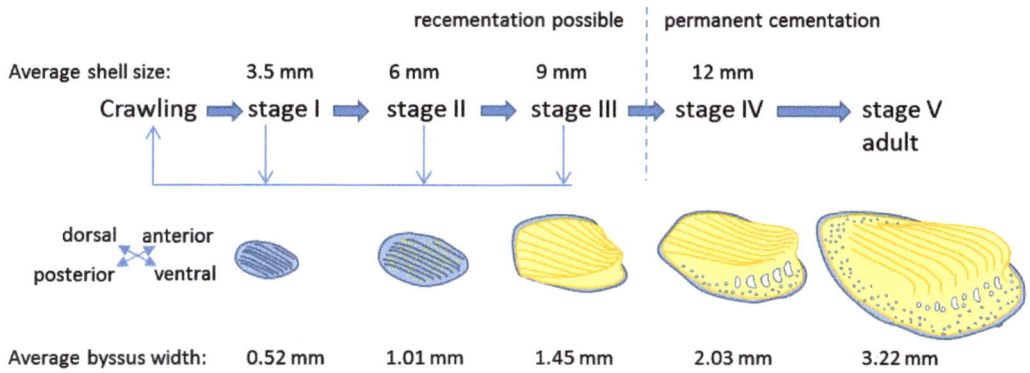

Figure 5 *Byssus formation during development in* Anomia chinensis. *See text for details. Adapted from Figures 4 and 8 in Yamaguchi.[9] The top part defines stages and summarizes animal mobility while the bottom part sketches the accompanying changes in byssus structure, where blue colours indicate organic matrix, while yellow represents mineralized areas.*

The large changes taking place during development reveal that the permanent byssus must perform as a jack of all trades, adapting to various load scenarios during growth of the animal. This observation has to be kept in mind when making functional conclusions based on structural data on adult individuals.

3.2 Organic Matrix

The calcified part of the byssus is interspersed with organic matrix that makes up less than 10 wt% of the byssus.[14] In spite of the relatively low quantity of organic material, decalcification by acid treatment leaves an intact organic framework, which reflects the shape of the original byssus. Pujol and coworkers have presented analysis of amino acid and saccharide composition in *Anomia simplex*.[14] They found that the dominant sugar was glucosaminoglycan that together with a small amount of galactosamine made up ~19% of the organic content of the byssus. In addition to these, smaller amounts of neutral sugars were also found: galactose, glucose, mannose, arabinose, xylose and fucose with galactose qualitatively appearing to be the predominant ones.

The relatively large quantity of glucosaminoglycan naturally leads one to speculate that chitin may be present in the byssus. However, Pujol et al. comment that they believe the glycosaminoglycan to be associated with glucoproteins rather than chitin.[14] More detailed investigations into this matter are under way in our laboratory. The protein composition is rich in acidic amino acids as one would expect for a highly calcified tissue.[14]

4 EXPERIMENTAL INVESTIGATIONS INTO THE MAGNESIUM CONTENT

The calcite polymorph of $CaCO_3$ is in many biomineralized tissues found to contain magnesium that substitutes for calcium in the crystal lattice.[15] Such substitutions alter the properties of the calcite phase, e.g. its solubility and mechanical properties. Previous work from our laboratory indicates that magnesium is not homogeneously distributed throughout the calcite byssus. To shed more light on this important aspect of byssus design, we investigated the magnesium content of the byssus in more detail as a part of the present work using a combination of synchrotron powder X-ray diffraction (SPXD) and scanning electron microscopy/energy dispersive X-ray spectroscopy (SEM/EDX).

4.1   Experimental

Specimens were obtained from the Marine Biological Laboratory (Woods Hole, MA, USA). Upon arrival they were either frozen (-18 °C) or kept alive in a custom built aquarium. The degree of Mg-substitution was determined either from synchrotron X-ray diffraction data by analysis of peak shifts or by energy dispersive X-ray spectroscopy (EDX) in the scanning electron microscope (SEM).

4.1.1   Synchrotron X-ray Diffraction. X-ray powder diffraction data were collected at the materials science beamline at the Swiss Light Source, Paul Scherrer Institute, Switzerland. Byssi were pulverized by an agate mortar and pestle and placed in a 0.3 mm diameter glass capillary. Diffraction data were collected using the beamline's Mythen detector that covers 120° 2θ by ~30000 individual detectors with an X-ray wavelength of 0.88629 Å.

Substitution of Ca by Mg in $CaCO_3$ led to a shift in peak positions towards larger diffraction angles due to a reduction of the lattice constants. To determine the magnesium distribution, the calcite (104) reflection was fitted using a model describing a diffraction contribution from a number of calcite phases containing a distribution of magnesium substitution. A total of 40 peaks placed at diffraction angles corresponding to 0-39 mol% magnesium substitution were used. The peak areas were fitted individually, but the peak widths were kept the same for all. By weighing the individual peak areas with their

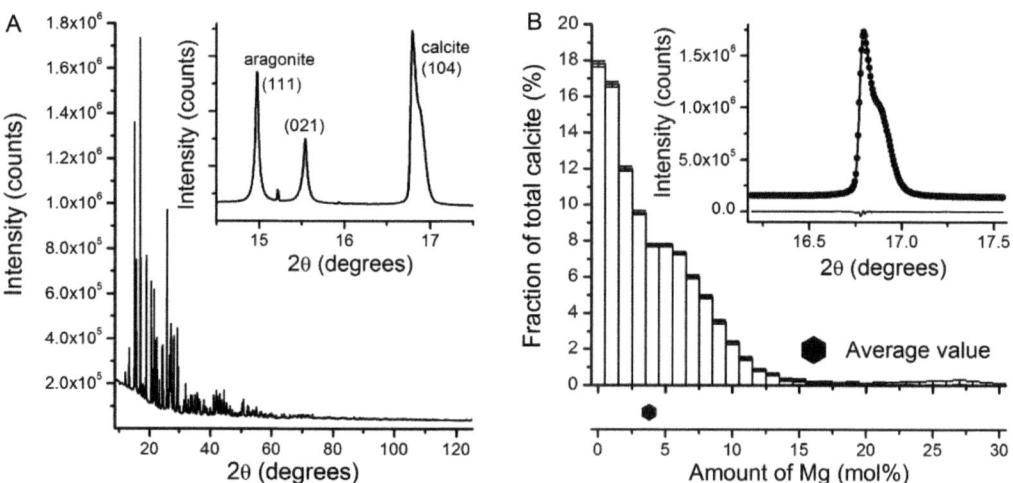

Figure 5   *Byssus crystallography investigated by synchrotron powder diffraction. (a) shows the full diffractogram with the inset focusing on a subset of the data featuring the calcite (104) reflection used in (b) to determine the Mg-distribution. In (b) the inset shows the fit (line) to the measured intensity distribution of the calcite (104) reflection while the main panel shows the ensuing histogram of Mg-substitution. The hexagonal dot below the histogram shows the average Mg-content.*

respective degree of Mg substitution and normalizing by the sum of areas, the relative amounts of magnesium substitution, i.e. the fraction of Mg, x, in $Mg_xCa_{1-x}CO_3$, were determined.

4.1.2 Element Mapping by EDX. A byssus detached from a rock was embedded in EpoFix (Struers, Denmark) and cut into slices using a water cooled diamond blade saw. The sections were subsequently polished using progressively finer diamond pastes finishing with 1/4 µm particle size paste. The slices were coated with carbon and imaged using back scattered electrons on a Maxim CamScan electron microscope using 20 kV electrons. EDX maps were collected overnight on the same instrument. The resulting intensity maps for Ca and Mg were turned into maps of at% Ca and Mg using standardless quantification. These at% maps were in turn used to calculate maps of the degree of magnesium substitution using at%(Mg)/(at%(Mg)+at%(Ca)). Only pixels with a significant amount of mineral, as determined by a threshold based on the Ca EDX signal, were included.

4.2 Results

Synchrotron X-ray powder diffraction, Figure 5a, confirmed that the byssi consist of both the aragonite and calcite polymorphs of $CaCO_3$.[9,11-13] The calcite (104) peak displays clear shoulders towards higher angles.[11] This means that there are several calcite phases present in the byssus with varying lattice constants. The main peak visible at ~16.8° 2θ corresponds to stoichiometric pure calcite meaning that the shoulders represent phases with smaller lattice constants. The most likely origin of this reduction in lattice constants is magnesium substitution.[15] By fitting the (104) peaks to a model consisting of peaks coming from a set of degrees of magnesium substitution, a distribution of the degree of magnesium substitution was obtained as displayed in Figure 5B. There is a large fraction of the calcite that contains no or very little magnesium followed by a step down towards a broad magnesium distribution that fades towards negligible contributions at about 14-15 mol% Mg. The average byssus calcite magnesium content was determined to be 3.79±0.04% mol% by this procedure. In the high-magnesium part of the byssus, the level of magnesium substitution corresponds to low-Mg calcite in sea urchin teeth that are renowned for their very high degree of magnesium substitution.[16] However, the amount is comparable to the values of 7.4-10.4 mol% found in oyster cement.[17]

The EDX maps are shown in Figure 6 (B-E) together with Back Scattered Electron (BSE) images (F-J). In BSE images, the signal is proportional to the local average atomic number, i.e. the higher the mineral content the higher the signal and the brighter the pixel in Figure 6. Figure 6A indicates the position of the sections shown in the remainder of the figure. The BSE images clearly show variation in the degree of mineralization in the lamellar part, seen particularly clearly in Figure 6F, where the contrast was further enhanced to emphasize this point. This contrast variation reflects the presence of organic sheets decorated with the herringbone arrangement of aragonite crystals.

Magnesium substitution is seen almost exclusively in the porous part reflecting the fact that magnesium only substitutes into the calcite and not the aragonite lattice. The degree of magnesium substitution is largest closest to the porous lamellar interface. Biogenic Mg-calcite displays stronger mechanical properties the Mg-free calcite.[18] This suggests that the inhomogeneous pattern of Mg-substitution may represent a functional adaptation: control over the degree of Mg-substitution leads to control over the local calcite mechanical properties and hence over the local material properties. Alternatively, Mg-substitution could reflect the composition of the mineralizing solution present during the byssus mineralization stage as has been suggested for oyster cement.[17] Which of these possibilities is at play in *Anomia* byssus is presently unclear. Resolving this issue will require a combination of detailed measurements of local mechanical properties combined with modelling of the multi length scale mechanical design of the byssus.

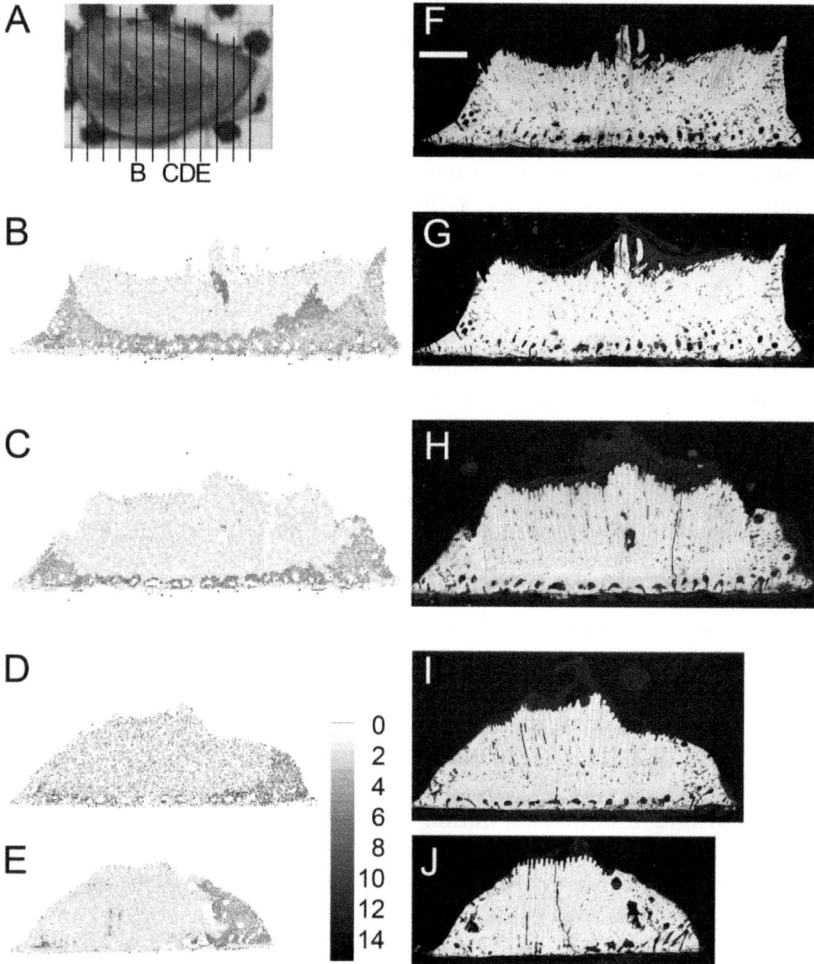

Figure 6 *Scanning electron microscopy and degree of Mg-substitution determined by EDX. Panel A shows a photograph of the samples with lines indicating where sections were cut. Small letters indicate the sections shown in the remaining images. Panels B-E is the degree of Mg-substitution (x in $Mg_xCa_{1-x}CO_3$) in the sections marked in panel A. The grey scale is defined by the scale bar to the right of panels D and E. Panels G-J shows backscatter electron images corresponding to the Mg-substitution maps in B-E. Panel F is the same section as panel G except that the contrast has been digitally increased to show the intensity variations in the lamellar portion more clearly.*

2 CONCLUSION

The byssus of *Anomia* displays a highly complex hierarchical structure. Intriguingly this joint-like structure is highly mineralized. Magnesium is inhomogeneously distributed throughout the calcitic porous part of the byssus as revealed in the present work by X-ray diffraction and EDX maps with the highest magnesium concentration found close to the porous/lamellar interface. The degree to which magnesium substitution is a part of the mechanical design of the byssus or caused by accidental incorporation during biomineralization is presently unclear. Investigations of mechanical properties of the byssus are currently underway in our laboratory to shed light on the relationship between the complex structure and the function of the byssus.

Acknowledgements

We thank the Human Frontiers Science Program and DANSCATT for funding, C. C. Broomell and J. H. Waite for helpful discussions, the Swiss Light Source for beam time, and Dr. Fabia Gozzo for kind assistance during the synchrotron measurements.

References

1   C. M. Yonge, *J. Mar. Biol. Ass. U.K.*, 1962, **42**, 113-125.
2   B. P. Lee, P. B. Messersmith, J. N. Israelachvili and J. H. Waite, *Annu. Rev. Mater. Res.*, 2011, **41**, 99 132.
3   J. H. Waite, *Biol. Rev.*, 1983, **58**, 209-231.
4   J. H. Waite, *Integr. Comp. Biol.*, 2002, **42**, 1172-1180.
5   J. H. Waite, N. Holten-Andersen, S. A. Jewhurst and C. J. Sun, *J. Adhesion*, 2005, **81**, 297-317.
6   J. H. Waite, H. C. Lichtenegger, G. D. Stucky and P. Hansma, *Biochemistry*, 2004, **43**, 7653-7662.
7   N. Holten-Andersen, M. J. Harrington, H. Birkedal, B. P. Lee, P. B. Messersmith, K. Y. C. Lee and J. H. Waite, *Proc. Natl. Acad. Sci. USA*, 2011, **108**, 2651-2655.
8   C. M. Yonge, *Phil. Trans. R. Soc. Lond. B*, 1977, **276**, 453-523.
9   K. Yamaguchi, *Marine Biology*, 1998, **132**, 651-661.
10 C. M. Yonge, *Nature*, 1957, **180**, 765-766.
11 J. R. Eltzholtz and H. Birkedal, *J. Adhesion*, 2009, **85**, 590-600.
12 R. S. Prezant, *American Malacological Bulletin*, 1984, **2**, 41-50.
13 J. R. Eltzholtz, M. Krogsgaard and H. Birkedal, in Structure-Property Relationships in Biomineralized and Biomimetic Composites, eds. D. Kisailus, L. Estroff, W. Landis, P. Zavattieri and H. S. Gupta, *Materials Research Society*, Warrendale, PA, 2009, pp. KK02-04.
14 J. P. Pujol, J. Bocquet, Y. Tiffon and M. Rolland, *Calc. Tiss. Res.*, 1970, **5**, 317-326.
15 F. T. Mackenzie, W. D. Bischoff, F. C. Bishop, M. Loijens, J. Schoonmaker and R. Wollast, in *Carbonates: Mineralogy and Chemistry*, ed. R. J. Reeder, Mineralogical Society of America, Washington DC, USA, 1983, pp. 97-144.
16 Y. Ma, B. Aichmayer, O. Paris, P. Fratzl, A. Meibom, R. A. Metzler, Y. Politi, L. Addadi, P. U. P. A. Gilbert and S. Weiner, *Proc. Natl. Acad. Sci. USA*, 2009, **106**, 6048-6053.
17 J. MacDonald, A. Freer and M. Cusack, *Marine Biology*, 2010, **157**, 2087-2095.
18 Y. Ma, S. R. Cohen, L. Addadi and S. Weiner, *Adv. Mater.*, 2008, **20**, 1555–1559.

QUALITATIVE AND QUANTITATIVE STUDY OF SPINY STARFISH (*MARTHASTERIAS GLACIALIS*) FOOTPRINTS USING ATOMIC FORCE MICROSCOPY

L.J. Higgins[1]* and A.S. Mostaert[1,2]

[1] School of Biology and Environmental Science, University College Dublin, Ireland.
[2] Conway Institute of Biomolecular and Biomedical Research, University College Dublin, Ireland.
*laila.higgins@ucdconnect.ie

1 INTRODUCTION

Starfish footprints are proteinaceous adhesive secretions produced by the individual tube feet of the animal and left behind on a surface after detachment.[1] Strong, but temporary adhesion is essential for this benthic invertebrate to withstand shear forces caused by water currents in its marine environment, and for successful feeding and defence against predators. The biological adhesive of starfish footprints adheres well to a wide variety of substrates, showing better affinity for high energy than for low energy surfaces, and greater adhesion on rough than smooth surfaces via increased contact area.[2,3,4] Their high performance in an aqueous environment has commercial potential, particularly for biomedical applications.

The study of biological adhesives is important for improving our understanding of how they function. This is of particular interest when considering adhesives produced by organisms in marine habitats.[5] Many marine organisms are currently under investigation for the purpose of either promoting adhesion for the development of aqueous-based biomimetic adhesives, or preventing adhesion towards exploring biofouling strategies. For example, the sandcastle worm adhesive (*Phragmatopoma californica*) has recently inspired improved approaches to the treatment of broken bones.[6]

The wide variety of marine organisms that produce biological adhesives is reflected in the high diversity of adhesion mechanisms that are currently known. Common to many is for the adhesive to be complex and heterogeneous, secreted by specialised organs evolved to mediate strong attachment. Sandcastle worms use a cement composed of oppositely charged polyelectrolytes which form a complex coacervation.[7] Mussels produce byssus threads, which are a complex interaction of polyphenolic proteins separated into an anchoring plaque and a collagenous thread.[8,9] Barnacle cement is composed of at least five proteins and undergoes a complex non-covalent curing process.[10] Limpets (various species) possess a mucous gel for locomotion, but change the characteristics to suit adhesion by altering the composition and concentration of the gel.[11] The parasitic marine flatworm, *Entobdella soleae,* secretes a proteinaceous adhesive containing insoluble, highly ordered amyloid fibrils providing both adhesive and cohesive strength to the adhesive.[12] Starfish (and other echinoderms) have also evolved a complex attachment mechanism using a strong, wet adhesive.[13] Unlike the examples provided, starfish are highly mobile, voracious predators, using their temporary adhesive effectively during locomotion and feeding.

Studies of the ultrastructure of starfish tube feet revealed two main secretory cell types responsible for creation of the adhesive.[13,14] The structure of the secreted adhesive was found to be surprisingly complex.[4] A homogeneous film is laid down first by specific secretory cells, and a spongy meshwork is laid on top by a second set of secretory cells.[4] A further type of secretory cell associated with detachment is thought to shed the uppermost layer of the tube foot cuticle[1], the so-called "fuzzy coat".[14] The fuzzy coat is believed to be then incorporated into the footprint to create a third adhesive layer which fills in the meshwork[1], but seems to break down when footprints are dried.[4] Overall, the adhesive secretion of starfish footprints is heterogeneous with little mixing, demonstrated by immunofluorescent staining[1] and lectin labelling[15] implying the presence of a complex microstructure within the footprint. One of the primary aims of this study was to investigate this structure at the micro- and nanoscale under native conditions.

Atomic force microscopy (AFM) is well suited to the challenges of obtaining high-resolution images and nanomechanical properties of starfish footprints; a soft, hydrated material, secreted in small quantities in a wet environment. The AFM technique is capable of obtaining *in situ*, nanoscale data with little to no sample preparation under native conditions. It has been used previously to study biological adhesives of marine organisms such as barnacle cyprids (temporary adhesive[9] and cement[16]), algae[17], diatoms[18], starfish[4] and monogenean flatworms[12], to name a few.

Here, AFM is used to investigate the nanoscale structure and nanomechanical properties of starfish adhesive footprints, from a previously unexplored species, *Marthasterias glacialis*. The goal was to provide an explanation for the underlying mechanical mechanism for adhesive strength of this underwater adhesive and to determine whether the heterogeneity seen in starfish footprints applies a functional micro- or nanostructure to the footprint. AFM enables combined mechanical and topographical measurements allowing for direct, *in situ* correlation of data, unlike other available techniques such as confocal z-stacks (topographical), SCIM (topographical) and nano-indentation (mechanical), making it very suitable for this experiment.

2 MATERIALS AND METHODS

2.1 Animal Collection and Maintenance

Adult spiny starfish (*Marthasterias glacialis*) were collected at low tide on the shore of Newquay, or from lobster pots set off the coast (County Clare, Ireland). They were kept in a recirculating marine aquarium (15° C) in artificial seawater (ASW) at 34 ppt (Instant Ocean, Sarrebourg, France), and fed on mussels (*Mytilus edulis*). All methodologies are in accordance with university and national legislation and guidelines.

2.2 Adhesive Footprint Collection

Footprints were collected by presenting an overturned starfish with a glass slide and allowing the tube feet to attach. Slides were first cleaned by sonication in 2% Decon 90 (Sigma, Ireland) for 15 minutes, then 100% ethanol for 15 minutes, before rinsing with ultrapure water (MilliQ) and drying under a stream of nitrogen gas. The reverse of the slide was kept dry so that when a tube foot adhered, the position of the resulting print could be marked accurately with a waterproof marker to allow location later. When each set of footprints was collected, the length of time the foot was attached to the surface was recorded. The footprints were gently rinsed and stored in filtered ASW before being imaged with AFM within 30 minutes of collection. Footprints were collected from the mid ray section of the starfish due to the preferential attachment using feet in this region.

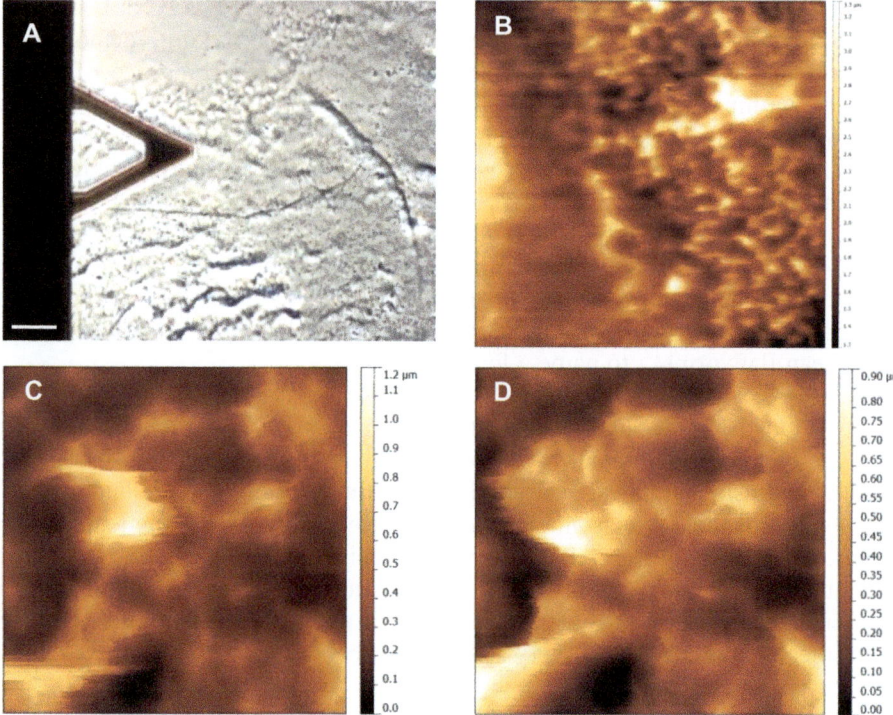

Figure 1   *AFM height images of meshwork layer in* M. glacialis *footprints. (A) Light micrograph of unstained footprint using phase contrast, with AFM cantilever approaching the sample surface shown. Scale bar = 50 μm. (B) AFM height image of the rough topography of the adhesive meshwork, and corresponding height scale. Scan size = 88 μm. (C) Detail of meshwork in the trace, and (D) retrace directions, showing the fibrous appearance of the topographical peaks. The structure of the meshwork is consistent, but offset due to the soft nature of the material. Scan size = 10 μm.*

2.3 Atomic Force Microscopy

All AFM data were obtained using a JPK Nanowizard II (JPK instruments, Berlin Germany) mounted on an inverted compound light microscope (Nikon Eclipse 80i). Measurements and calibrations were performed in filtered ASW using tip C of DNP cantilevers (Bruker, Santa Barbara, USA) with nominal spring constants ranging from 0.12 to 0.48 N/m and resonant frequencies ranging from 40 to 75 kHz, in air. Average tip radius was 20 nm. The footprints were located using the inverted light microscope with the help of the pen markings on the back of the slides. Optical images of the footprints were collected using phase contrast before imaging with AFM. Qualitative measurements were made in amplitude modulation mode with a scan rate of 0.7 - 1.5 Hz. The cantilever sensitivity was calibrated using an inverse optical lever sensitivity (InvOLS) measured from the slope of the constant compliance region in the imaging medium (ASW) against a hard substrate (glass). The spring constant was determined using the method of Hutter and Becchoefer[19] (thermal calibration), Q factor was determined by fitting thermal spectrum of the cantilever to a single harmonic oscillation (SHO) model. Imaging settings were optimised for each sample and adjusted as appropriate to maintain high quality imaging. In order to investigate the influence of imaging force on the footprints, the imaging setpoint was set to the free amplitude of the cantilever and gradually decreased until evidence of imaging contrast was observed. Scans were then performed sequentially, keeping the imaging parameters constant, but lowering the imaging setpoint by ~5 % for each scan. The lower limit was determined by the observation of feedback artefacts when the setpoint was a small percentage of the cantilever's free amplitude (A_0).

Energy dissipation was calculated using the phase (φ) and amplitude (A) images of an amplitude modulation mode scan, recording A_0 in advance, 50 μm above the substrate. This allows for measurement of the *in situ* relative energy dissipation map of the sample which can be related to the samples stiffness and/or adhesiveness. Imaging parameters were kept constant for all images of the same footprint, and as close as possible between footprints while still maintaining imaging quality. ED was calculated in Gwyddion[20] according to the following equation:

$$\frac{1}{2}\frac{k\omega}{Q}\left[A_0 A \sin\varphi - A^2\right] \tag{1}$$

where k is the spring constant, ω is the resonance frequency of the cantilever, and Q is the quality factor.[21,22]

3 RESULTS

3.1 Imaging Using Height Measurements

Consistent with previous descriptions of starfish adhesive footprints,[4] those of *M. glacialis* were circular and colourless (Figure 1A). The footprints were easily located using phase contrast and could be precisely positioned beneath the AFM cantilever (Figure 1A). There was a variation in thickness across the print, and scan regions needed to be chosen avoiding regions that were too thick to image. Nevertheless, care was taken to select scan site from the centre and near the edge of the footprints.

The meshwork was visible, however the distinction between it and the underlying homogeneous film was not as clear (Figure 1B), nor was the globular nanostructure visible, compared to previous study of starfish footprints.[4] The meshwork appears to be one continuous soft layer of variable height. The structure of this layer appears to be composed of fibrous peaks and troughs that create the meshwork (Figure 1C). The retrace (Figure 1D) was consistent with the trace image (Figure 1C), though offset due to the softness of the material. Height range in general was typically 200 to 500 nm, but was noted to exceed 2 µm on occasion. Such high variations in topography combined with the soft nature of the samples can often cause difficulties for optimising imaging conditions. To overcome these problems the scan size was reduced and the scan rate was increased to 1.5 Hz. Slower scan rates may have allowed time for this material to adhere to the cantilever.

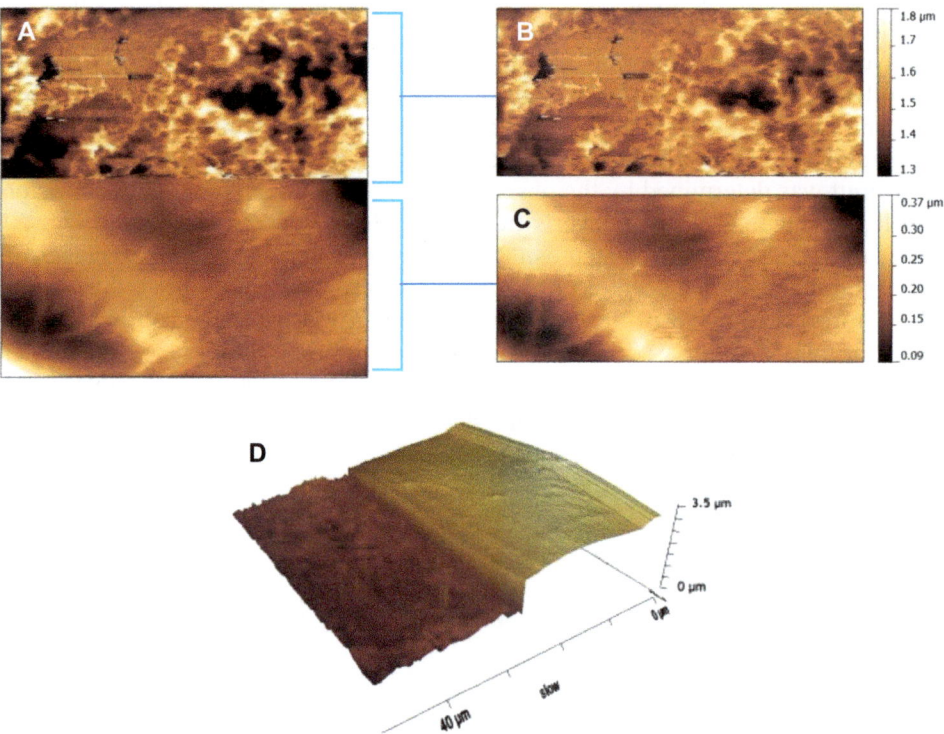

Figure 2   *Height images of* M. glacialis *adhesive footprints where (A) the setpoint was halved part way through the scan, revealing the meshwork (B) beneath the soft, gel-like matrix (C). Scan size =50 µm². (D) Tilted rendering showing an approximate 800 nm height difference between the meshwork and gel-like matrix layers for this image. Scan size = 50 µm².*

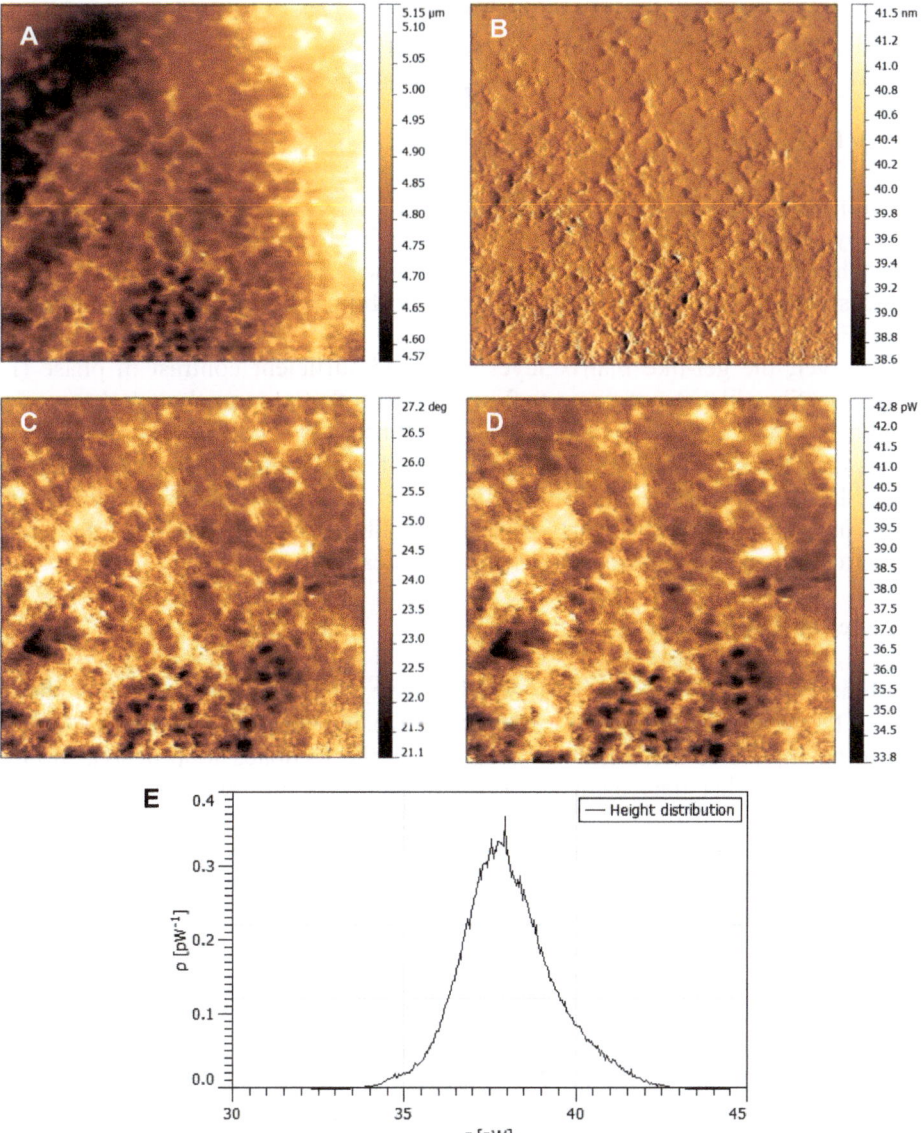

Figure 3 *The meshwork layer of* M. glacialis *footprints showing the (A) height, (B) amplitude, (C) phase and (D) energy dissipation. Scan size = 50 μm. The height distribution for energy dissipation (E) shows a continuous gradient rather than a bimodal separation, indicating a lack of significant contrast within the scan.*

It became evident during the course of these AFM measurements that there was a third layer above the meshwork which was a soft gel-like material with interwoven fibrous structures (Figure 2A-C). This material exhibited low roughness and provided little resistance when a vertical force was applied. Fibres were visible in the height image (Figure 2B), but were not observed to directly correlate with the meshwork underneath (Figure 2C). These fibres were not always observed in the upper layer.

3.2 Energy Dissipation

Energy dissipation was calculated to be in the range of 30 to 50 pW, with highest energy dissipated by the meshwork (Figure 3). This may be due to crosstalk from the high contrast in topography (Figure 3A). The distribution of values for energy dissipation falls neatly into a Gaussian distribution, showing no significant bimodal separation of features within the layers (Figure 3E).

However, the gel-like matrix layer exhibited sufficient contrast in phase (Figure 4C) to identify fibre-like structure up to 2 μm in width that are mostly not observed in the height image (Figure 4A). These fibres demonstrated significantly higher energy dissipation than the surrounding material (Figure 4D), as indicated by the second, minor peak in the histogram (Figure 3E). Since this contrast did not correlate with topography, a difference in mechanical properties between the fibres and the surrounding matrix as opposed to a difference in adhesive properties is indicated.

3.3 Influence of Applied Force

In amplitude modulation AFM (AM AFM) mode, the setpoint represents the target amplitude (TA) of the cantilevers. Therefore decreasing the setpoint increases the force required to be applied to the cantilever resulting in the feedback loop decreasing the tip-sample distance. Further decreasing the setpoint increases the imaging force resulting in the tip interacting more strongly and typically closer to the surface of the sample. In Figure 5, imaging began at a height where there was no interaction with the sample and then the setpoint was successively lowered. Image contrast was evident with a setpoint of 104 nm. Figure 5A shows the surface to be relatively smooth with a height of 1.2 μm with no visible meshwork. The setpoint was then decreased to 80 nm (Figure 5B), where a meshwork pattern beneath the smooth layer began to emerge as the tip interacted more strongly with the sample, and the maximum height difference increased to 1.4 μm. The meshwork became more pronounced when the setpoint was further lowered to 70 nm (Figure 5C) and 60 nm (Figure 5D), however it began showing signs of sample compression. With further decrease to 50 nm (Figure 5E), a reduction in the height difference and imaging artefacts (feedback oscillations) were observed. Imaging was unstable at 45 nm with all other imaging parameters held constant.

4 DISCUSSION

The starfish adhesive system is considered to be one of many complex adhesive systems that are common in nature. Comparable to other species of echinoderm studied,[23] *M. glacialis* produces a multi-layered secretion composed of a homogeneous film, a topographically rough meshwork and a soft gel-like matrix with an embedded web of long

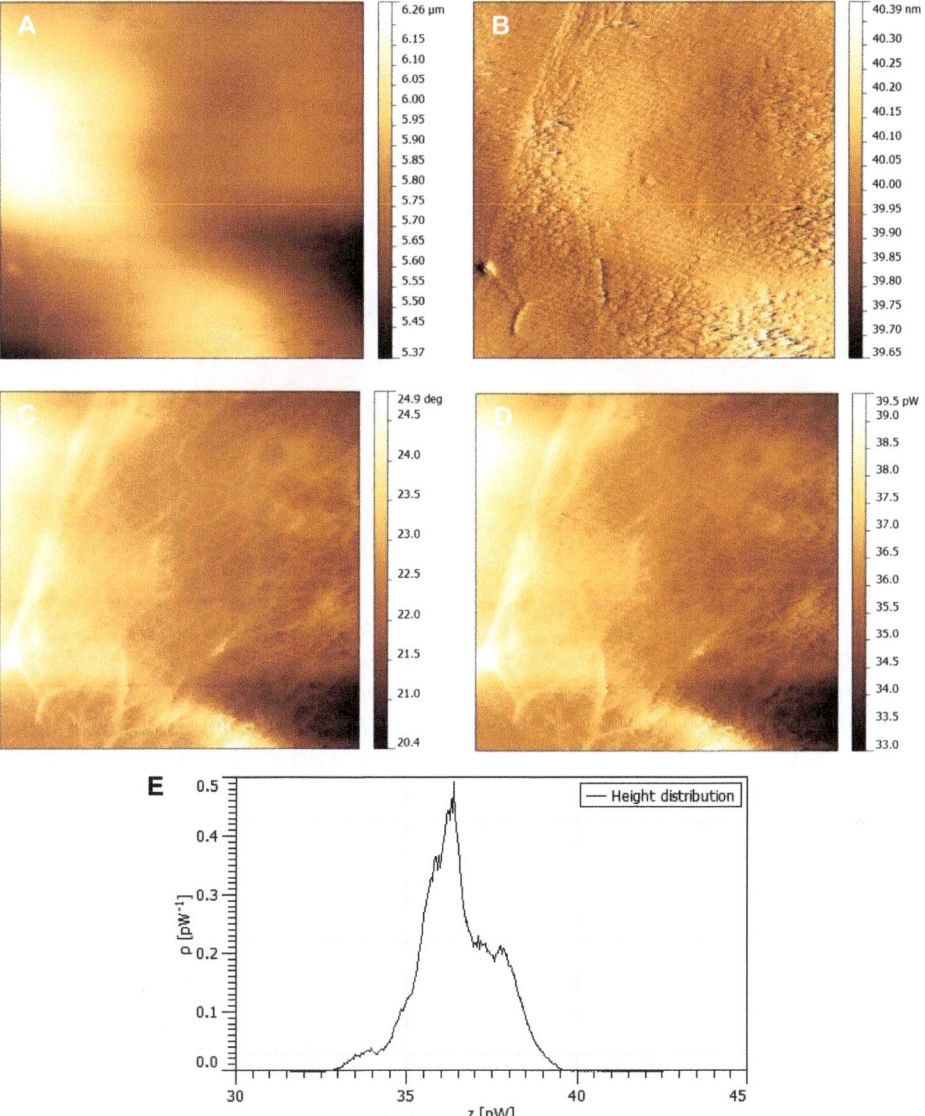

Figure 4 *The upper matrix layer of* M. glacialis *footprints showing the (A) height, (B) amplitude, (C) phase and (D) energy dissipation. Scan size = 50 μm. The fibres visible in the phase (C) and energy dissipation (D) signals are almost invisible in the height (A) and amplitude (B), indicating a contrast in mechanical properties. The height distribution for energy dissipation (E) shows a secondary peak of higher energy, indicating a significant contrast within the scan.*

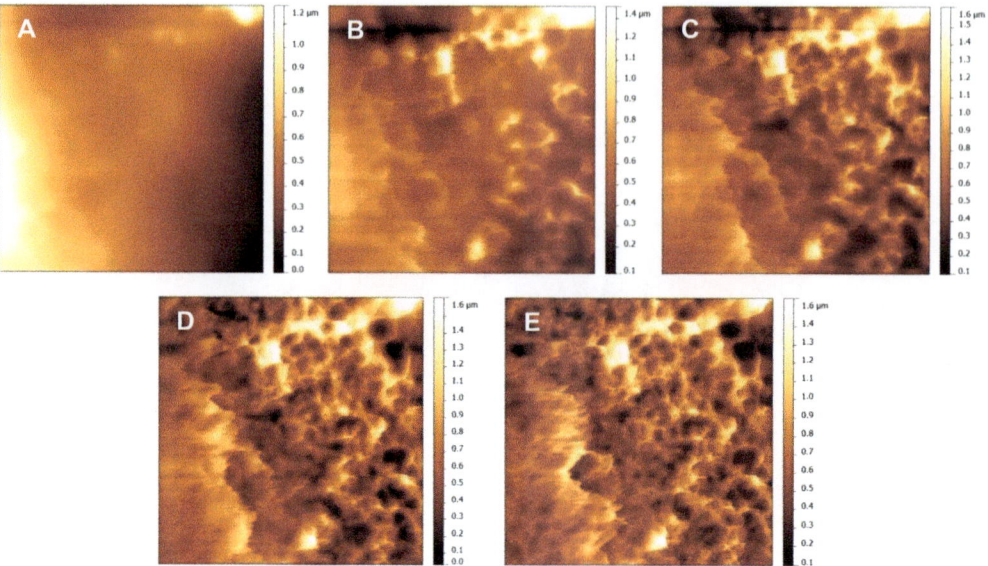

Figure 5   *AFM height images taken in contact mode showing z-scale force increase on M. glacialis adhesive footprints. Scan = 50 μm. (A) Smooth layer, no meshwork (104 nm). (B) Top of the meshwork beginning to appear (80 nm). (C) Meshwork more pronounced (70 nm). (D) Left side of the image shows signs of compression (60 nm). (E) Feedback oscillation artefacts indicate that the force exerted on the sample is at or has exceeded the viscoelastic properties of the material (50 nm). Scan size = 30 μm².*

interweaved fibres enabling the starfish to maintain a strong hold on the seafloor.

In this study, the distinction between the previously described homogeneous film and meshwork[4] was not clear. These two layers showed little contrast in mechanical properties based on calculated energy dissipation, and in the height images they appeared to form one continuous layer. This is likely a result of crosstalk from the rough topography. There was also a notable absence of globular material previously reported in *Asterias rubens*.[4] Here the film and the meshwork exhibited greater fluidity and appeared to adopt a fibrous structure. It is possible that footprints imaged in their native conditions maintain a higher water content, making them more difficult to image clearly. Reflecting this, height values for the meshwork were greater in *M. glacialis* than those measured for *A. rubens*. However the height difference and lack of globular structures could reflect species specificity in the adhesive secretions between *M. glacialis* and *A. rubens*, as has been shown in other echinoderms.[23]

Within the secretion there was, however, a distinct viscosity contrast between the gel-like matrix and the meshwork, indicating a difference in composition and function. Previously, the uppermost layer, the gel-like matrix, has only been characterised using transmission electron microscopy, where its appearance alternated between loose electron-lucent material and granular material of two types: heterogeneous electron dense material and homogenous material of medium electron density.[1,4,13,23] Cryo-scanning electron microscope images show a fibrillar meshwork deposited on a priming film.[23] Using AFM, the soft matrix was successfully imaged and shown to possess an internal structure of fibres with a mechanical contrast to the surrounding material, an ultrastructure which is consistent with a fibre reinforced composite.[24] The uppermost peaks of the meshwork also demonstrated a fibrous appearance, though the authors cannot comment on whether or not these features are homologous. The gel-like layer could represent part of the fuzzy coat which is incorporated into the footprint during detachment.[1,14] Its response to force applied by the cantilever supports the postulation that this layer breaks down in response to drying.[4]

The intact presence of the third layer provides further evidence that the meshwork is not an artefact of detachment, but a structural component within the footprint, as proposed by Hennebert *et al.* (2008),[4] where it was shown that the meshwork was unaffected by various sample preparation methods, and that the hexagonal pattern was likely derived from the relative position of secretory cell pores on the surface of the tube foot disc. The meshwork material has viscoelastic properties represented by its response to increasing force caused by the tip. A degree of elasticity within the footprint material could provide extra load-bearing capabilities, adding to the high tenacity exhibited by many starfish species.[13,23]

The presence of a smooth, fibrous soft layer on top of the adhesive footprint could explain how starfish footprints retain high adhesion despite the use of a deadhesive secretion. The gel-like matrix could act as insulation, protecting the underlying meshwork from both the effects of the deadhesive and the external environment post-detachment. The softness of the material could be a result of the detachment process. Most adhesives are very tacky when first secreted, but quickly lose their adhesive properties as they cure and harden.[6,9,10,24] The pliability of the top layer of starfish secretions allows access to the layers beneath. In contrast, only the exposed cement of barnacles is available for investigation, because the internal structure is insulated by a much stiffer outside layer than presented here.[9] It also emphasises the importance of investigating bioadhesives under physiological conditions, as thus far, this is the first presentation of the *in situ* characteristics of the gel-like matrix intact.

An adhesive needs to be sticky, so that it bonds two surfaces together and requires energy to separate them.[24] It also needs to adhere to itself (cohesion) so as not to create a weak point.[26] Adhesive strength depends on the characteristics of the adhesive, the surfaces involved, and the conditions under which the adhesive is to operate.[25] In biological systems, there is an added layer of complexity owing to the development of specialised structures via evolutionary processes that affect the mechanical properties of the adhered bodies (the secreting organs).[24] For example the mutable collagen within the stem of the starfish tube foot provides a variable stiffness,[26,27,28] and therefore a variable adhesive strength or tenacity.[28]

Energy dissipation has been demonstrated in another biological system, the cell, where it has been shown to reveal a stiffness contrast not visible in a typical height image, which showed the presence of actin geodesic nodes in the substructure of the plasma membrane.[29] It also highlighted the stiffness contrast in the plasma membrane created by the organelles' relative positions within the plasma membrane. Energy dissipation therefore shows great potential for imaging in soft biological systems in real time.[30]

Here, where the topography demonstrated high contrast there is a possibility that crosstalk between the height and phase signals may have obscured the measurement of structure of the underlying meshwork. Where the topography was smoother in the softer top layer, the microstructure was resolved in detail. Despite some limitations of AFM highlighted in this work, it has been shown to be a valuable technique for the investigation of a complex biological adhesive under native conditions.

Acknowledgements

We sincerely thank Dr Jason Kilpatrick for helpful advice on the analysis and interpretation of AFM data. L.H. was supported by a Demonstratorship from the School of Biology and Environmental Science, University College Dublin.

References

1   P., Flammang, A., Van Cauwenberge, H., Alexandre, and M. Jangoux, *J. Exp. Biol.*, 1998, **201**, 2383.
2   L.A., Thomas, and C.O. Hermans, *Biol. Bull.*, 1985, **169,** 675.
3   R., Santos, S., Gorb, V., Jamar, and P., Flammang, *J. Exp. Biol.*, 2005, **208**, 2555.
4   E., Hennebert, P., Viville, R., Lazzaroni, and P., Flammang, *J. Struct. Biol.*, 2008, **164**, 108.
5   K., Kamino *Mar. Biotechnol.* , 2008, **10**, 111.
6   H., Shao, K.N. Bachus, and R.J., Stewart, *Macromol. Biosci.*, 2009, **9**, 464.
7   R.J., Stewart, J.C., Weaver, D.E., Morse, J.H., Waite, *J. Exp. Biol.*, 2004, **207**, 4727.
8   J.H., Waite, *Biol. Rev.*, 1983, **58**, 209.
9   H.G. Silverman and F.F., Roberto *Mar. Biotechnol.*, 2007, **9**, 661.
10 I.Y. Phang, N. Aldred, X.Y. Ling, J. Huskens, A.S. Clare and G.J. Vansco, *J. R. Soc. Interface*, 2010, **7**, 285.
11 A.M. Smith, *Intergr. Comp. Biol.*, 2002, **42**, 1164.
12 A.S. Mostaert, R. Crockett, G. Kearn, I. Cherny, E. Gazit, L. C. Serpell and S. P., Jarvis, *Arch. Hist. Cytol.*, 2010, **72**, 199.
13 P. Flammang, in *Biological Adhesives*, ed. By A.M. Smith and J.A. Callow, Berlin, Heidelberg, Springer-Verlag, 2006, pp. 183–206.
14 P. Flammang, S. Demeulenaere, and M. Jangoux, *Biol. Bull.*, 1994, **187**, 35.
15 E. Hennebert, R. Wattiez and P. Flammang, *Mar. Biotechnol.*, 2010, **13**, 484.
16 I.Y. Phang, N. Aldred, A.S. Clare, J.A. Callow and G.J. Vansco *Biofouling*, 2006, **22**, 245.
17 J.A. Callow S.A. Crawford, M.J. Higgins, P. Mulvaney and R. Wetherbee, *Planta*, 2000, **211**, 641.
18 T.M. Dugdale R. Dagastine A. Chiovitti P. Mulvaney and R. Wetherbee, *Biophys. J.*, 2005, **89**, 4252.
19 J.L. Hutter and J. Becchoefer, *Rev. Sci. Instrum.*, 1993, **64**, 1868.
20 D. Nečas and P. Klapetek, *Cent. Eur. J. Phys.* 2012, **10**,181-188. http://gwyddion.net/

21 J.P. Cleveland, B. Anczykowski and V.B., Elings, *Applied Physics Letters*, 1998, **72**, 2613.

22 B. Anczykowski, B. Gotsmann, H. Fuchs, J.P. Cleveland and V.B. Elings, *Appl. Surf. Sci.* 1999, **40**, 376.

23 R. Santos, E. Hennebert, A.V. Coelho and P. Flammang, in *Functional surfaces inn biology*, ed. S.N. Gorb, 2009, pp 9-41

24 H. Lee, N.F. Scherer and P.B. Messersmith, *PNAS*, 2006, **103**, 12999.

25 C. Gay, *Integr. Comp. Biol.*, 2002, **42**, 1123.

26 R.S. McCurley and W.M. Kier, *Biol. Bull.*, 1995, **188**, 209.25

27 R. Santos, D. Haesaerts, M. Jangoux and P. Flammang, *J. Exp. Biol.*, 2005, **208**, 2277.

28 E. Hennebert, D. Haesaerts, P. Dubois and P. Flammang, *J. Exp. Biol.*, 2010, **213**, 1162.

29 P. Maguire, J.I. Kilpatrick, G. Kelly, P.J. Prendergast, V.A. Campbell, B.C. O'Connell and S.P. Jarvis, *HFSP Journal*, 2007, **1**, 181.

30 R. García, R. Magerle, and R. Perez, Nature materials, 2007, **6**, 405.

MODELLING OF BIOMIMETIC SYSTEMS

HOW GEOMETRY AFFECTS THE ADHESION OF GECKO-LIKE ADHESIVES

M. Röhrig*[1], M. Thiel[2], S. Bundschuh[3], M. Worgull[1] and H. Hölscher[1]

[1] Institute of Microstructure Technology (IMT), Karlsruhe Institute of Technology (KIT), Eggenstein-Leopoldshafen, Germany
[2] Nanoscribe GmbH, Eggenstein-Leopoldshafen, Germany
[3] Institute for Applied Materials (IAM), Karlsruhe Institute of Technology (KIT), Eggenstein-Leopoldshafen, Germany
*michael.roehrig@kit.edu

1 INTRODUCTION

The phenomenal adhesive properties of gecko toes have been extensively investigated for a long time, resulting into the development of novel adhesive tapes.[1-4] Besides the strong adhesion to nearly any substrate, their outstanding attachment system allows geckos to detach within milliseconds.[5] This remarkable combination of strong attachment as well as rapid and easy detachmentoriginates from the hierarchical design of delicate hairs covering the lamellae that are crossing the toe-pads. These hair are called setae and are about 4 μm in diameter and 100 μm in length. Finally, they branch into hundreds of tiny endings, the about 200 nm wide spatulae (*Gekko gecko*).[6] Due to this hierarchical design, geckos achieve very intimate contact to flat and even to relatively rough surfaces, enabling them to climb walls and ceilings only with the help of van-der-Waals interactions.[7] Consequently, the fabrication of synthetic gecko adhesives mimicking these phenomenal nano- and microstructures is pursued by numerous groups all over the world.[8-21]

Currently, the soft molding technique is the most often used approach for the fabrication of synthetic gecko adhesives.[8,11,24] Usually, materials applicable for soft molding like PDMS or polyurethane are cast into etched silicon wafers or SU-8 templates. However, the soft molding technique implies some drawbacks. Due to the demolding process and the need of complex designs for gecko-like adhesives, it is essentially restricted to soft materials. Furthermore, demolding is a delicate process were the mold might be destroyed by accident or on purpose.[9,25] Considering that the observed adhesion depends very strongly on structure-design parameters like pillar dimension, aspect ratio and tip shape, mold inserts are an inflexible and elaborate approach for design studies, because a new mask and/or mold has to be manufactured for every parameter variation.[26-28] So far, several geometry induced effects, like the improvement of adhesion by hierarchical structuring, could only be analyzed for soft materials and not for stiff materials at the relevant nanometer scale.[14,25]

In order to cope with these issues, we present 3D direct laser writing (DLW) for the rapid prototyping of hierarchical gecko-inspired surfaces with elastic modulus and relevant length scales matching the gecko's toe-pads very closely. In contrast to previous studies we perform normal adhesion experiments on those gecko-inspired structures. The obtained

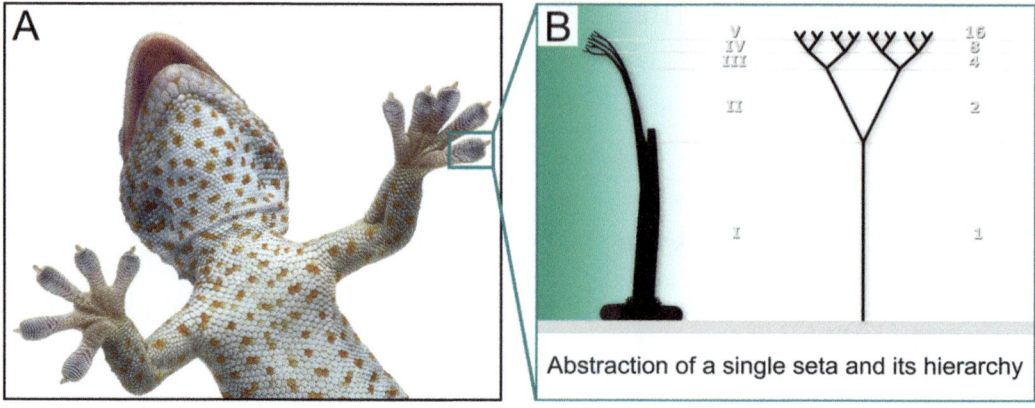

Figure 1   *A) A bottom view of a Tokay gecko (Gekko gecko).[22] The toe pad of geckos is typically separated into distinct lamellae. B) These lamellae are covered with millions of delicate hairs, the so-called setae. The setae are multi-scale hierarchical structures, which are split up to four times. The tiniest endings are the so-called spatulae with a width of about 200 nm (adapted from Batal[23]).*

results show that hierarchical structures are indeed favorable for stiff materials on the nanoscale as expected from numerous theoretical studies.[28-33] In contrast to molding techniques, 3D direct laser writing offers the quick realization of design concepts that are neither restricted to demoldable designs nor limited by any mold fabrication technology. This gives us the highest flexibility in creating gecko-mimicking surfaces.

2 RESULTS AND DISCUSSION

2.1 Robust Design

In order to ensure robust adhesion, the design of gecko-mimicking structures needs immense optimization to prevent functional failure of the adhesive tape like fibrillar bunching, for example. In the following, existing models are discussed, expanded and adjusted for this work.[27,28,30-32,34-45]

At first, we consider an adhesive structure consisting of fibers with radius R, length L and inter-fiber distance $2D$. Assuming further that the Johnson-Kendall-Roberts (JKR) theory can be applied, the pull-off force required to detach a spherical tip from a flat and infinite stiff substrate is given by

$$F_c = \frac{3}{2}\pi RW,\tag{1}$$

where W represents the work of adhesion.[46]

2.1.1 Failure by Fiber Fracture. In order to increase the adhesion by the principle of contact splitting or enhanced compliance, the fiber radius should be selected as small as possible.[5,47] However, Spolenak et al.[27] proposes, that if the fiber radius became too small, the axial stress σ_c may exceed the theoretical fracture strength σ_f resulting in fiber fracture. Hence, the correlation

$$\sigma_c = \frac{F_c}{\pi R^2} \le \sigma_f \qquad (2)$$

limits the useful fiber radius R:

$$R \ge \frac{3W}{2\sigma_f} \approx 15\frac{W}{E_f}, \qquad (3)$$

where the theoretical fracture strength is approximated by $E_f/10$, with E_f being the elastic modulus of the fiber.[48]

2.1.2 Failure by Exceeding the Ideal Contact Strength. The ideal contact strength σ_{th} that is transmitted through the actual contact area at the instant of tensile instability, is the upper limitation for the contact strength:[27]

$$\sigma_c = \frac{F_c}{\pi r_c^2} \le \sigma_{th}, \qquad (4)$$

with r_c being the contact radius at the instance of pull-off. Considering the JKR theory and a rigid contacting surface, r_c can be expressed as

$$r_c = \left(\frac{9\pi W R^2 \left(1 - v_f^2 \right)}{8E_f} \right)^{1/3}. \qquad (5)$$

Here, v_f is the Poisson ratio of the fiber. Combining eqs. (4) and (5), the second lower limit for the fiber radius R can be extracted:[41]

$$R \ge \frac{8s^3 E_f^2}{3\pi^2 W^2 \left(1 - v_f^2 \right)} \qquad (6)$$

2.1.3 Failure by Crack Propagation. Due to surface roughness, the contact within the contact zone of fiber and substrate is never perfect. The resulting defects in the interface can be modeled as cracks, in order to use fracture mechanics to solve this problem.[30] Tensile loading of the fiber may cause instable crack propagation within the interface, leading to reduced adhesion. Considering Griffith's criterion, the critical load for the instable propagation of an existing crack of the length 2a is for an elastic solid[49]

$$\sigma_g = \sqrt{\frac{16G\gamma}{\pi a (\kappa + 1)}}, \qquad (7)$$

with G being the shear modulus, γ being the surface energy and κ describing a coefficient depending on state of plain stress or strain respectively. Since this equation is only valid for a crack in an elastic solid, it has to be adopted to describe a crack in the interface of fiber and substrate. While a crack propagates in a brittle solid, the following energies are contributing: the strain energy U, the surface energy of the crack S and the potential energy of the external forces P. The Griffith's criterion supposes, that strain energy is released while crack propagation. For an instable propagation of the crack, the released strain energy has to overcome the required surface energy, hence, resulting in a reduction of the total potential energy. Consequently, the Griffith's criterion for a crack to propagate by *da* calculates to

$$\frac{d}{da}(U + S + P) \leq 0. \tag{8}$$

Assuming the model of a disk in a field of uniaxial tensile stress, for a Griffith crack which is *2a* in length, the sum of strain energy and potential energy is[50]

$$U + P = -\frac{\pi(\kappa+1)}{8G}\sigma_t^2 a^2. \tag{9}$$

The surface energy of a crack in the interface between fiber and substrate is

$$S = 2a(\gamma_f + \gamma_s) \tag{10}$$

where γ_f and γ_s are the surface energies of the fiber and the substrate. Combining eqs. (8), (9) and (10) leads to the instability criterion

$$\frac{d}{da}\left[-\frac{\pi(\kappa+1)}{4G}\sigma_t^2 a^2 + 2a(\gamma_f + \gamma_s)\right] \leq 0. \tag{11}$$

Providing that the energy released by extending the crack by *da* overcomes the required surface energy, instable crack propagation occurs. Hence, the critical Griffith load for a crack within the interface of an adhesive contact between fiber and substrate is

$$\sigma_{fs} = \sqrt{\frac{E(\gamma_f+\gamma_s)}{\pi a}}, \tag{12}$$

with $G = \frac{E}{2+2v}$ for isotropic materials and assuming a state of plain strain ($\kappa = 3 - 4v$). In case of ideal contact strength being below the critical Griffith load, instable crack propagation is prevented:

$$\sigma_{th} \leq \sigma_{fs}. \tag{13}$$

Hence, in order to tolerate a defect which is $^1/_4\,R$ wide and existing in the interface between fiber and substrate, the fiber radius hast to fulfill the condition

$$R \leq \frac{4E_f(\gamma_f+\gamma_s)}{\pi\sigma_{th}^2}. \tag{14}$$

2.1.4 Failure by Bunching. In a fibrillar array, attractive van-der-Waals forces may cause bunching of neighbouring fibers, resulting in a decreased compliance of the structures. For a given distance 2D between two fibers, a critical length L exists above which bunching occurs. In the following, the anti-bunching condition for fibers is proposed. For simplicity, the anti-bunching condition is calculated for a quadrangular cross-section of the fibers and a quadrangular pattern.

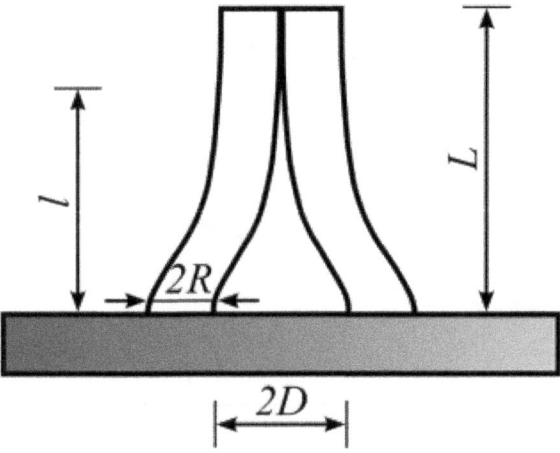

Figure 2 *For a given distance 2D between two fibers, bunching occurs if the length of the fibers L exceeds a critical value. Van-der-Waals forces forward the bunching, whereas restoring forces of the deformed fibers contribute to the separation of the fibers. Adapted from Hui et al.[34]*

Considering two bunched fibers touching along the distance *L-l* (Figure 2), the equilibrium is dominated by two contributing energies: the adhesion along the interface and the potential energy of the deformed fibers. In order to separate two bunched fibers, the energy

$$2U_{adh} = -2 \int_{l}^{L} \gamma_f (2R) \, dy = -2\gamma_f (2R)(L - l) \qquad (15)$$

is required. In order to model the potential energy of the fibers, we assume a clamped beam of length *L* guided by a floating bearing at *l* and a linear bearing at its end (Figure 3). The bearing reaction F_b is

$$F_b = 12\frac{DE_f I}{l^3}, \qquad (16)$$

with $I = \frac{(2R)^4}{12}$ being the moment of inertia of a fiber with quadrangular cross section, and *D* representing the deflection of the fiber. Hence, the potential energy of the deformed fiber is

$$U_{pot} = \int_{x=0}^{D} F_b x dx = \frac{6E_f I D^2}{l^3}. \qquad (17)$$

The fibers will tend to adopt the state with the lowest energy. If no minimum in the effective potential is existing in the interval $0 \le l/L$, the fibers do not bunch. Summing eqs. (15) and (17), the effective potential is

$$U_{eff} = U_{adh} + U_{pot} = -\gamma_f (2R)(L - l) + \frac{E_f (2R)^4 D^2}{2l^3}. \qquad (18)$$

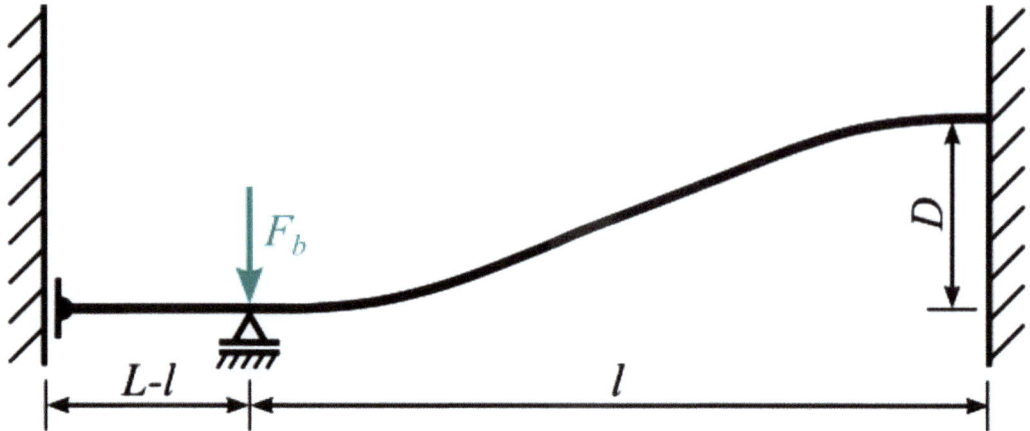

Figure 3   *The bunched fiber is modelled as clamped beam guided by a float bearing at L and a linear bearing at its end.*

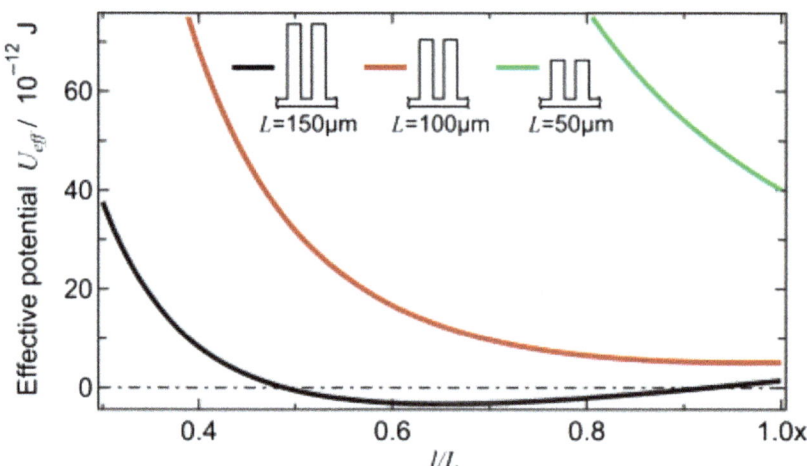

Figure 4   *By adjusting the length L of the fibers, a minimum in the effective potential is avoided, preventing bunching of the fibers. (γ_f = 33 mJ m^{-2}, E_f = 4 GPa, 2R = 5 µm, D = 2 µm)*

In the following, we substitute l/L by χ for simplicity. In order to prevent a minimum of the effective potential in the interval $0 \leq \chi \leq 1$, the condition

$$U'_{eff}(\chi) = \gamma_f (2R)L - \frac{3E_f(2R)^4 D^2}{2\chi^4 L^3} \neq 0, \quad 0 \leq \chi \leq 1 \tag{19}$$

has to be fulfilled. Setting $\chi = 1$ (critical limit) and solving equation (18) defines the maximum fiber length L for the prevention of bunching:

$$L \leq \left(\frac{3}{2}\right)^{1/4} \sqrt{D} \left(\frac{(2R)^3 E_f}{\gamma_f}\right)^{1/4}. \tag{20}$$

The distance between two fibers is *2D* and can be expressed as a function of the effective area fraction φ and the maximum achievable area fraction φ_{max} for the chosen fibrillar pattern.[30]

$$2D = 2R\left(\sqrt{\frac{\varphi_{max}}{\varphi}} - 1\right), 0 \leq \varphi \leq \varphi_{max}. \tag{21}$$

Combining equations (20) and (21) leads to the final anti-bunching condition

$$L \leq \left(\frac{3}{2}\right)^{1/4} \sqrt{R\left(\sqrt{\frac{\varphi_{max}}{\varphi}} - 1\right)} \left(\frac{(2R)^3 E_f}{\gamma_f}\right)^{1/4}. \tag{22}$$

2.1.5 Maximizing Work of Adhesion. Considering only van-der-Waals energies and neglecting others, the work of adhesion of a single-level fibrillar array can be expressed as

$$W = \varphi \Delta\gamma = \varphi(\gamma_f + \gamma_s - \gamma_{fs}) \tag{23}$$

where γ_f, γ_s and γ_{fs} represent the surface energies of fiber and substrate, as well as the energy of the fiber-substrate interface. For a multi-level, hierarchical fibrillar array, it is reasonable to add the elastic strain energy to the work of adhesion.[30] Assuming a cylindrical fiber of primary length L, the fiber is elongated by the ideal contact strength by

$$\Delta L = L \frac{\sigma_{th}}{E_f}. \tag{24}$$

The provided work $W_{diss} = \sigma_{th}\Delta L$ to elongate the fiber is dissipated by material inherent damping when the fiber detaches.[38] In order to calculate the work of adhesion for a hierarchical, e.g. two-level, fibrillar array, this dissipation has to be added to the van-der-Waals contribution:

$$W_2 = \varphi_1(\Delta\gamma + W_{diss}) = \varphi_1\left(\gamma_f + \gamma_s - \gamma_{fs} + \frac{L\sigma_{th}^2}{E_f}\right). \tag{25}$$

In order to increase the robustness of the adhesive structures, the work of adhesion has to be maximized.

2.1.6 Iterative Dimensioning. According to the previously described limits, the dimensions of our gecko-mimicking arrays were calculated with an iterative procedure described in Figure 5. The dimensions chosen for our study are listed in Table 1.

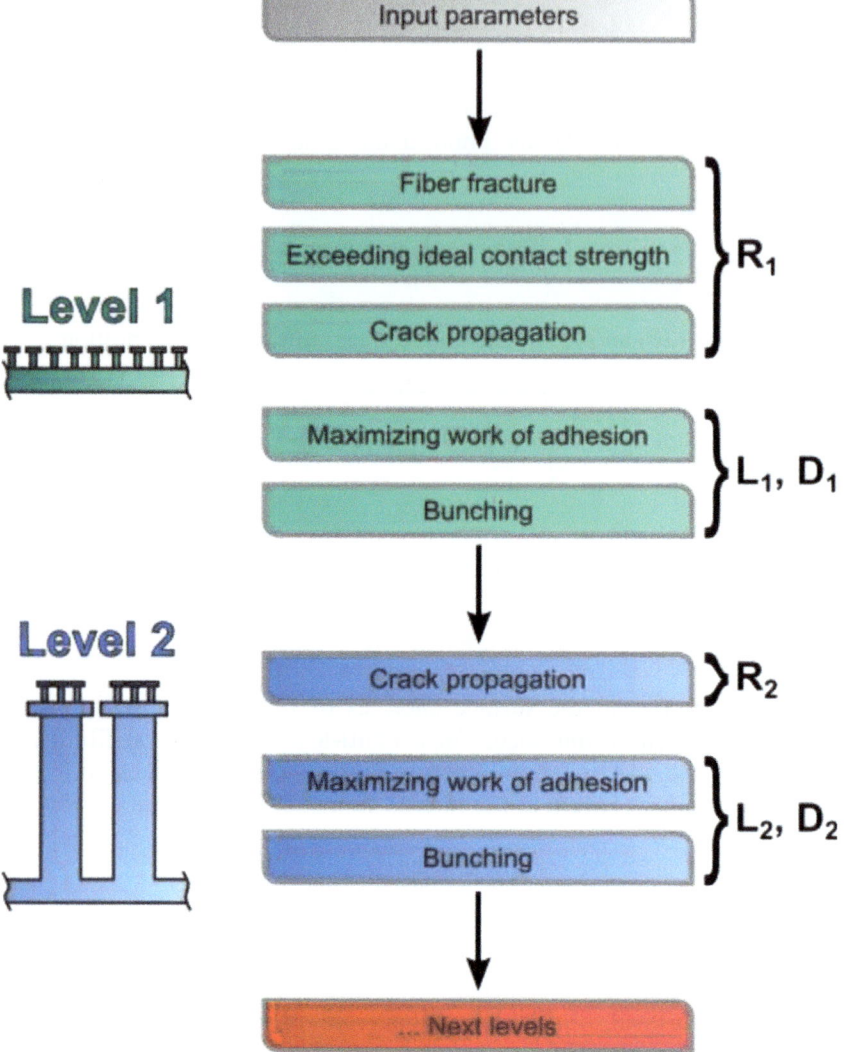

Figure 5   *The flow-chat shows the algorithm to calculate the fiber dimensions iteratively for every hierarchy level.*

2.2 3D Direct Laser Writing

Applying 3D direct laser writing, we fabricated gecko-inspired nano- and microstructures.[51] Using an acrylic based negative tone resist (IP-G 780) for highest resolution in multiphoton absorption enables us to design arbitrary structures down to the nanoscale.[52,53] Additionally, the elastic modulus of the resist ($E_{IP-G\ 780} \approx 4$ GPa) is closer to the ones of the gecko ($E_{gecko} = 1 - 4$ GPa) than softer materials like poly(dimethylsiloxane) ($E_{PDMS\ Sylgard\ 184} = 2.6$ MPa), polyurethane ($E_{PUR\ ST-1060} = 3$ MPa), or polyvinylsiloxane ($E_{PVS} = 3$ MPa) which are often used for the fabrication of dry adhesives.[5,25,29,53-55] Consequently, 3D direct laser writing is the perfect tool for a design study of hierarchical gecko-mimicking structures. Depending on their body weight, different gecko species

Table 1 *Dimensions of all fabricated gecko-mimicking arrays (all values are in μm). The values emphasize the variation of the parameters h_1 and p_1 leading to a decreased density (arrays 2, 5, and 8) and aspect ratio (arrays 3, 6, and 9). The values marked with an * correspond to arrays with mushroom-shaped tips.*

parameter	1	2	3	4	5	6	7	8	9
a_1	0.5/0.56*	0.5/0.56*	0.5/0.57*	5.6/5.6*	5.6/5.6*	5.6/5.6*	5.6/5.7*	5.6/5.6*	5.6/5.6*
h_1	2.25/2.1*	2.25/2.27*	1.8/1.9*	24.5/27.9*	24.5/27.9*	16.8/15.9*	24.5/27.5*	24.5/28.4*	16.8/13.0*
p_1	1.1/1.1*	1.4/1.4*	1.1/1.1*	9.0/9.0*	11.0/11.0*	9.0/9.0*	9.0/9.0*	11.0/11.0*	9.0/9.0*
r_1	0/0.9*	0/0.9*	0/0.9*	0/7.6*	0/7.6*	0/7.6*	0/7.6*	0/7.6*	0/7.6*
t_1	0/0.9*	0/0.9*	0/1.0*	0/2.0*	0/2.1*	0/1.35*	0/2.0*	0/2.1*	0/1.2*
a_2	–	–	–	–	–	–	0.5/0.57*	0.5/0.6*	0.65/0.65*
h_2	–	–	–	–	–	–	1.8/2.7*	2.3/2.7*	1.6/1.0*
p_2	–	–	–	–	–	–	1.1/1.1*	1.4/1.4*	1.1/1.1*
r_2	–	–	–	–	–	–	0/0.9*	0/0.9*	0/0.75*
t_2	–	–	–	–	–	–	0/1.4*	0/1.4*	0/0.15*

array number

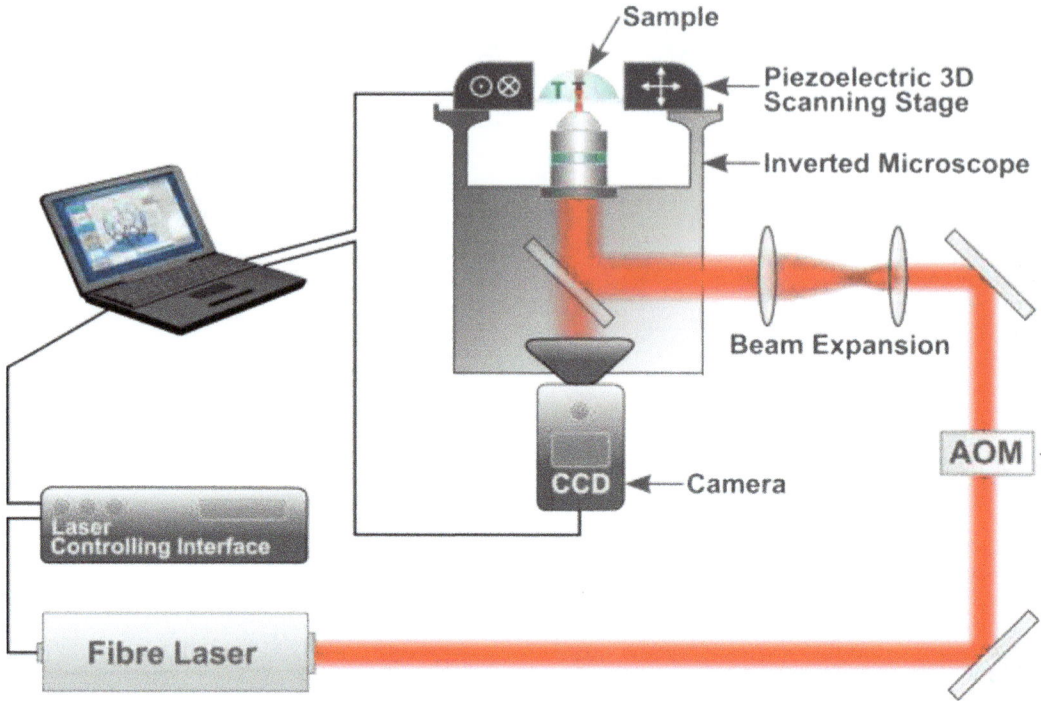

Figure 6   *Schematic of the 3D direct laser setup used for this study. The beam of the ultra-short pulsed fibre laser is focussed into the photoresist by a high numerical aperture objective. The control of the laser power with an acousto-optic modulator (AOM) allows scaling of the voxel. By moving the piezoelectric scanning stage, the sample position can be shifted relatively to the fixed focal position, enabling the writing of arbitrary paths into the material.*

evolved two-fold (e.g., *Anolis carolinensis*), three-fold (e.g., *Hemidactylus turcicus turcicus*) or even quaternary branched (e.g., *Gekko gecko*) hierarchical structures.[1,2,56] In order to show the basic characteristics and advantages of 3D direct laser writing for biomimetic adhesives, we restricted ourselves to two-fold hierarchical structures. Some of the results have been previously reported,[51] here we present a more detailed analysis.

2.2.1 Technology. 3D direct laser writing is a rapid prototyping technique based on multiphoton absorption, enabling the fabrication of arbitrary 3D nanostructures in suitable photoresists like IP-G 780 (Figure 6).[52] The photoresist is perfectly transparent for the laser light since the one-photon energy lies below the absorption edge of the material for the chosen wavelength. However, by tightly focussing the light of the ultrashort-pulsed laser, in the focal volume the intensity is high enough to expose the photoresist by multiphoton absorption. Within this small volumetric pixel ("voxel"), the absorption causes a chemical and/or physical modification of the photosensitive material. Using a developer bath, the unexposed regions are removed after the writing.[57]

2.2.2 Elastic Modulus of the IP-G 780 Resist. The gecko's setae are consisting of beta-keratine with an elastic modulus of $1 - 4$ GPa.[5,29,53] It is only the hierarchical design of these setae that leads to a very low effective modulus near 100 kPa.[5] Building these

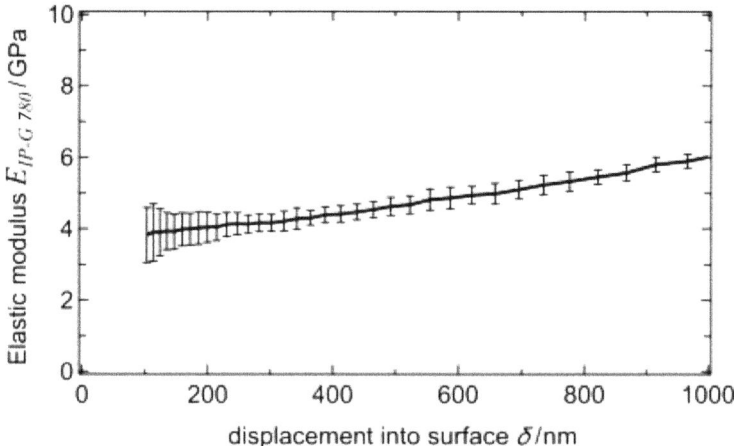

Figure 7 *Nanoindentation experiments were performed to characterize the elastic modulus of the used photo resist IP-G 780. The figure shows the calculated elastic modulus from CSM measurements. A common observation for soft and compliant thin films on harder and stiffer substrates, also seen here, is the rise of the elastic modulus with increasing indentation depth. A common assumption for this is that up to 10% of the total film thickness, the measured values correspond to the real film properties. Hence, the elastic modulus of IP-G 780 is approximately 4 GPa, measured between 100 nm and 200 nm indentation depth.*

setae out of a stiff material has many advantages, like the prevention of structural collapse despite high density and aspect ratio of the fibers or improved self-cleaning.[53] Hence, a stiff material was chosen to accomplish gecko-mimicking structures by 3D direct laser writing. The elastic modulus of the used photoresist IP-G 780 was determined by nanoindentation (Nano Indenter XP; Agilent Technologies, Santa Clara, CA, USA). The indentation experiments were conducted using the dynamic contact module (DCM) with a Berkovich tip. The samples were indented with a constant strain rate of 0.05 s^{-1} to a depth of 1000 nm. Besides the load and displacement data, the instrument provides information on the contact stiffness continuously during the loading process via a superimposed displacement oscillation of 2 nm at 45 Hz (continous stiffness method, CSM).[58] By means of the measurement in Figure 7, the elastic modulus is determined to 4 GPa.

2.2.3 Fabrication of Gecko-Inspired Arrays. For the fabrication of the presented gecko-inspired structures, a coverslip served as substrate for the photoresist. First, this coverslip was cleaned with isopropyl alcohol as well as acetone and blown-off with nitrogen afterwards. This preparation was followed by spin-coating a 200 nm thin layer of SU-8 (1. 500 rpm for 10 seconds with acceleration of 100 rpm/second. 2. 2600 rpm for 59 seconds with acceleration of 300 rpm/second). During the following processing, this thin layer of SU-8 on top of the coverslip ensures reliable bonding of the photoresist to the substrate. After soft baking (100°C, 2 minutes), the SU-8 layer was exposed completely (6 minutes with a 36 W UV-source). Afterwards the chosen photoresist IP-G 780 was dispensed on top by using a pipette. Baking the compound for 90 minutes at 100°C secures the post-exposure bake of the SU-8 layer and soft-bake of the IP-G 780 resist. Subsequently, the sample was inserted into the laser lithography system. After completion

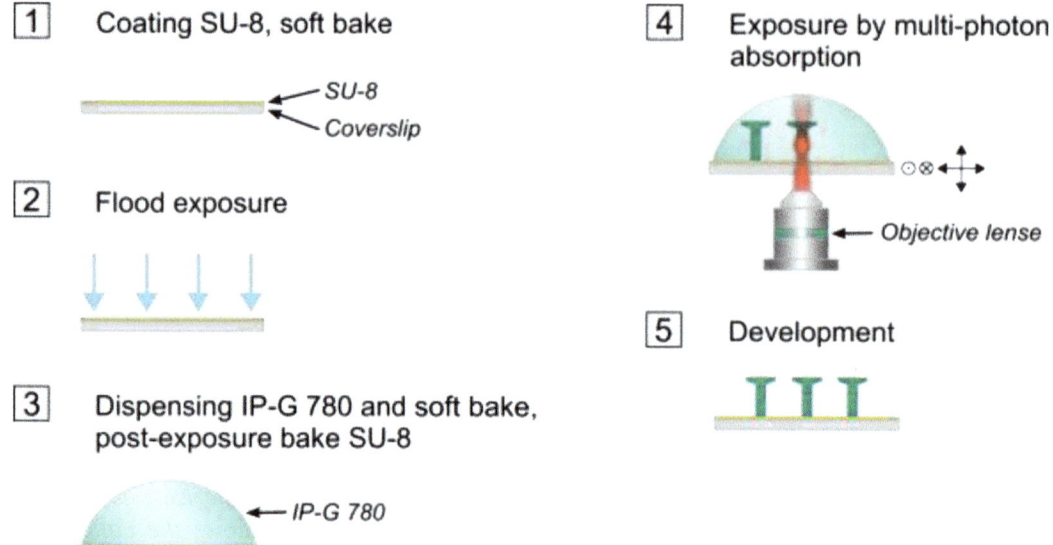

Figure 8   *Schematic of the five processing steps for the fabrication of gecko-inspired nano- and microstructures by 3D direct laser writing.*

of writing, the structures were developed for 30 minutes in a PGMEA developer bath, followed by a 5 minutes cleaning in isopropyl alcohol and drying by nitrogen. The performed processing steps are summarized in Figure 8.

Figure 9 and Table 1 summarize the structures and dimensions fabricated and investigated in this study. Depending on the actual set of parameters, the resulting quadrangular arrays have an edge length between 68 μm and 82 μm. The structure's dimensions were verified by scanning electron microscopy at tilt angles of 0° and 70°. The width of the square pillars ranges from 500 nm to 5 μm. The periodicities of the pillars are between 1.8 and 2.8 times their width, whereupon the aspect ratios reach values up to 4.5.

2.3 Adhesion Measurements

In the past, friction measurements were performed on hierarchical structures with a similar elastic modulus, however, the structures appear to be stochastic and their normal adhesion was not investigated.[59-61] In the following, we therefore perform a normal adhesion analysis and investigate the impact of design variations on the adhesion of gecko-inspired structures with an elastic modulus close to the gecko's setae. Due to the comparable small areas of the available arrays, the adhesion measurements had to be performed by atomic force microscopy (AFM).[62]

2.3.1 Procedure. Force-versus-distance measurements allow for the investigation of surface interactions and properties, i.e. adhesion force, separation energy, and compliance (Figure 10A). The tip apex radius of conventional AFM cantilevers, however, is in the range of nanometers, and much smaller than the smallest lateral dimension of the fabricated samples (Figure 10B). Therefore, a spherical silica particle of about 20 μm in diameter was mounted on a tipless AFM cantilever (Figure 10C).[63] This set-up also eliminates the need for a complex alignment control, required for a flat probe.[64]

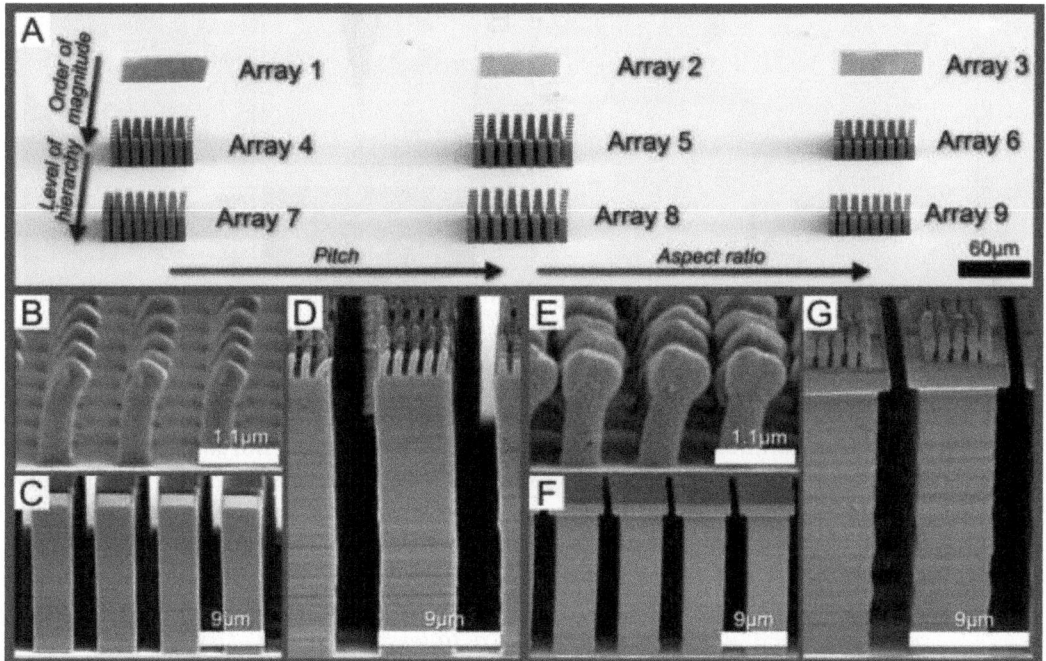

Figure 9 *A) By using 3D direct laser writing, several arrays of gecko-inspired structures were fabricated. The arrays in the left column are optimized with regard to their aspect ratio and density. The density is reduced in the middle column and the aspect ratio is reduced in the right column. In addition, we combined the smaller and larger pillars to get a two-fold hierarchy (array 7, 8, and 9) and changed the tip shapes. B) SEM image of array 1, which contains small single level structures with a width of only 500 nm. C) The SEM image shows array 4, which contains pillars of 5 μm width. The dimensions are 10 times larger than they are in array 1. D) Array 7 consists of the array 1 on top of array 4. E), F) and G) show corresponding arrays with mushroom-shaped tips. All scale bars refer to the periodicity of the arrays (parameter p_1 in Table 1).*

So-called force maps were measured for investigating the adhesion of the fabricated structures (Figure 11). Each force map covered a minimum area of 25 μm × 25 μm and contained at least 1024 (= 32 × 32) force-versus-distance curves. For analyzing the adhesion, at least four force maps were measured for a preload of 100 nN, 500 nN, 1 μN, 2 μN, 3 μN, 4 μN, 5 μN, 6 μN and 6.7 μN. The averaged adhesion force was additionally referenced to the cross-sectional area of the spherical silica particle in order to identify the contact strength.

The velocity during approach and retraction of the cantilever has to be chosen carefully, since visco-elastic effects are influencing the pull-off force.[65] In this sense, the adhesion of selected arrays was measured as a function of the velocity. As expected, the adhesion has an upper boundary value that is reached for velocities fasterthan 12 μm/s.

Analyzing the contact of a flat IP-G layer with the used silica probe allows us to estimate the contact area during the measurements. By the inspection of the Tabor coefficient for the given contact, it can be decided whether the JKR model or the DMT model has to be applied.[46,66,67] The JKR model leads to more realistic description of the

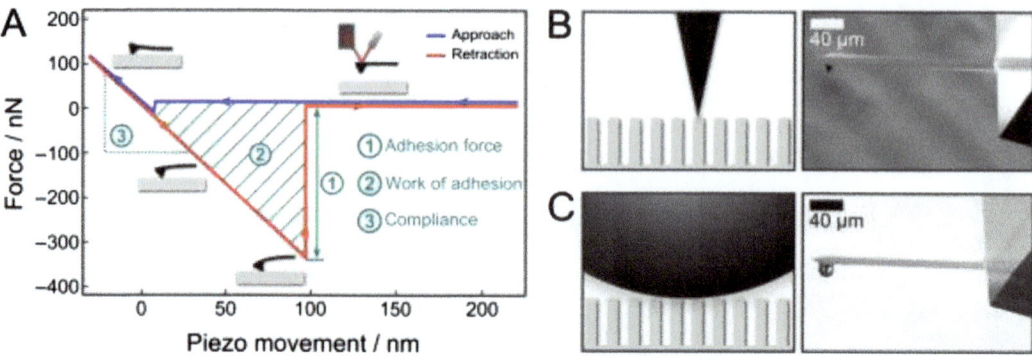

Figure 10 *A) A force-versus-distance measurement comprises the approach and retraction of an AFM cantilever. If the cantilever is close enough to the sample surface, it feels the interaction forces. The resulting force-versus-distance plot reveals the adhesion force, the work of adhesion and the compliance. B) Since the tip apex radius of conventional cantilevers is much smaller than the smallest lateral dimensions of the fabricated samples, they are not suitable for our samples. C) Instead of using sharp tips, the adhesion measurements were performed with a spherical silica particle attached to a tipless cantilever. Using a spherical particle eliminates complex alignment control.*

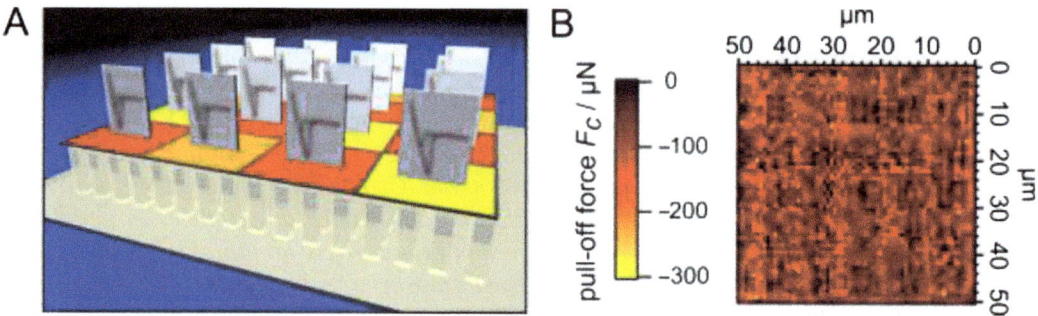

Figure 11 *A) For measuring a force map, the sample surfaces is divided into several pixels. On every pixel a force-versus-distance curve is measured. After analyzing the measurements, the adhesion is shown as a color contrast in the force map. B) As an example, a force map of 25 µm x 25 µm is shown. The force map is based on 64 x 64 force-versus-distance measurements.*

contact between large and soft solids, whereas the DMT model is more suitable for small, hard solids.[68] Supposing the radius of curvature $C_f = 10$ µm, the elastic modulus $E_f = 70$ GPa, the Poisson's ratio $v_f = 0.17$ for the silica probe and $C_s = \infty$, $E_s = 4$ GPa and $v_s = 0.22$ for the IP-G layer and assuming a surface energy of $\gamma = \gamma_f = \gamma_s = 45 \cdot 10^{-3}$ Nm^{-1} the Tabor coefficient is

$$\mu = \frac{h}{z_0} = \left(\frac{\gamma^2 C^*}{z_0^3 E^{*2}}\right)^{1/3} \approx 47.5, \tag{26}$$

where h is the neck height around the contact zone and Z_0 is the equilibrium separation of the atoms with a typical value of 3 Å.[66] C^* represents the radius of curvature of the silica probe and E^* the reduced elastic modulus as defined by:[46]

$$\frac{1}{C^*} = \frac{1}{C_f} + \frac{1}{C_s} \text{ and } \frac{1}{E^*} = \frac{1-v_f^2}{E_f} + \frac{1-v_s^2}{E_s} \tag{27}$$

Since $\mu > 1$, the JKR model is valid for the given contact.[68] The real contact radius between a sphere and a flat surface is given by[46]

$$r^3 = \frac{3C^*}{4E^*}\left(F + 3\pi WC^* + \sqrt{6\pi WFC^* + (3\pi WC^*)^2}\right). \tag{28}$$

where r is the real contact radius, W the work of adhesion and F is the external force. With $W = \gamma_f + \gamma_s = 2\gamma$ this leads to a contact area with a diameter between 700 nm and 820 nm for the given preload range. The contact area at pull off is[69]

$$r_c = 0.63\left(\frac{6\pi WC^{*2}}{E^*}\right)^{1/3} \approx 350 nm. \tag{29}$$

The largest real contact area during a force-versus-distance measurement is, therefore, at the maximum compressive load and is expected to be < 1 % than the cross sectional area of the spherical probe.

A significantly reduced real contact area compared to the apparent contact area is characteristic of rough contacts. The simplest rough surface could be imagined as a surface uniformly covered by asperities which have all the same radius of curvature C and the same height (Figure 13A).[70] The center of all asperities in Figure 13A is at the position $z = 0$, however, real surfaces are usually randomly rough. The model of Greenwood and Williamson[71] enhances the previously described model with random height of the asperities (Figure 13B). The center of each asperity n is displaced by Δd_n from the mean plane of center, whereby the distribution of Δd_n is Gaussian.

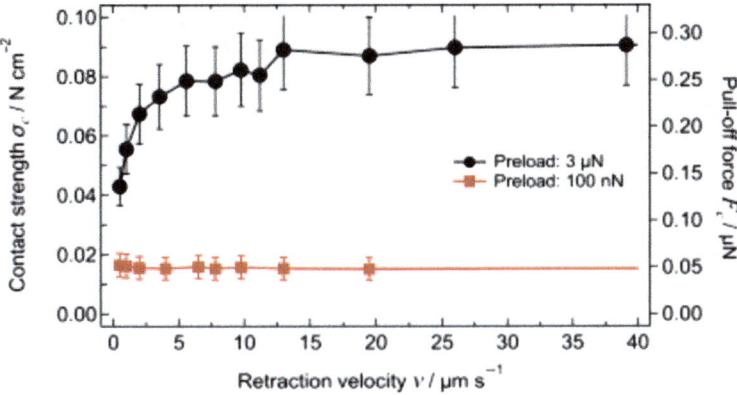

Figure 12 *The adhesion has been measured as a function of the retraction velocity. For velocities faster than 12 µm/s, the adhesion reaches its upper boundary value.*

While performing the adhesion measurements, a spherical probe (diameter: 20 μm) was pressed into the sample with a given preload. The relative position d of the spherical probe while performing the measurements is exemplarily shown in Figure 13C (measurement of array 3 with mushroom-shaped tips; preload: 6 μN). The corresponding deviations of the center mean position Δd_n is Gaussian distributed. Assigning a summit with radius of curvature $C = 10$ μm to each height position d_n leads to a rough surface topography with the root-mean-squared roughness of $R_q = 8.7$ μm. The distribution of Δz_m, the deviation of the height from its mean height, is shifted in a way that is characteristic of

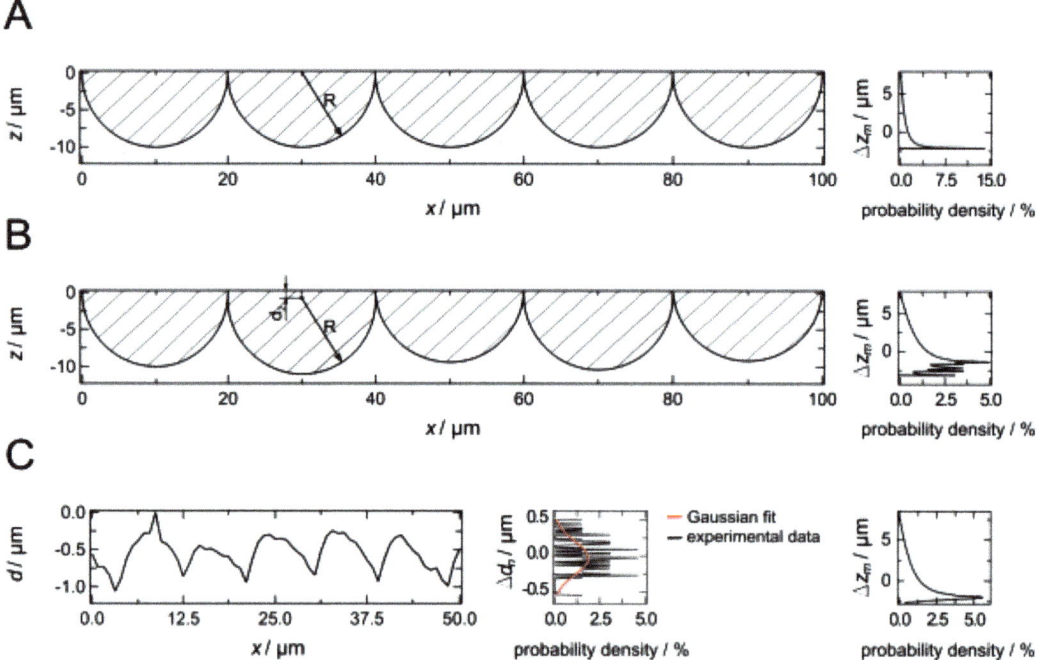

Figure 13 *A) The simplest roughness could be imagined as uniformly distributed asperities which have all the same radius of curvature C and the same height. Since the center of the asperities are all at the same height instead of Gaussian distributed, such a surface does not describe a roughness in the proper meaning of the word, but rather a profile. The height probability density referring to the mean height of the profile is shown on the right. B) According to Greenwood and Williamson[71], a roughness can be modeled by asperities whose summits have all the same radius of curvature C, but with random height. The corresponding height probability density becomes Gaussian distributed. C) In the adhesion measurements that were performed, the spherical probe touches the sample at different heights. The relative heights are exemplarily plotted as the relative height d vs. the lateral coordinate x. The distribution of the relative height is Gaussian, as required by the model of Greenwood and Williamson. Assigning a summit with radius of curvature C = 10 μm to each height position leads to a rough surface topography with a skewed height probability density that is typically for rough surfaces with high peaks. Hence, the adhesion measurement with a spherical probe imitates the contact with a rough surface.*

rough surfaces with high peaks. Thus, the performed adhesion measurements approximate the contact of the sample with a micro-rough surface.

2.3.2 Results. In order to investigate how geometry affects the adhesion of gecko-like adhesives, contact strength versus preload plots were recorded. For the plots shown in Figure 14, at least four force maps were averaged. As a reference, all plots contain the adhesion vs. preload curve (crosses) of a flat sample surface that has been fabricated by 3D direct laser writing as well. As expected, the measured adhesion of this flat reference sample is very low and exhibits basically no preload dependency.

Compared to the cylindrical pillars, the pillars with mushroom-shaped tip show an improvement of the adhesive properties in all our measurements. This mushroom-shape does not only increase contact area, it also improves the stress distribution in the contact area during detachment as recently discussed by Carbone *et al.*[72] Figure 14A depicts the increased adhesion of the mushroom-shaped tips (solid circles) compared to the unstructured pillars (open circles) by the example of the two-folded hierarchical structures. The improvement is seen for all preloads.

In Figure 14B, the impact of density and aspect ratio is exemplified for the smaller pillars with mushroom-shaped tips (solid squares). A reduction of the density of about 20 % (solid stars) causes a negative effect comparable to a decrease in aspect ratio of 10 % (solid diamonds). This leads to an average loss in adhesion of 40 % for less density, and of even 50 % in case of a decreased aspect ratio. These slight modifications of the structures already demonstrate the strong influence of the design on the actual adhesion. However, one of the challenges in fabricating dry adhesives is the prevention of a structural collapse while increasing density and aspect ratio in order to improve the adhesion.[26,45]

In Figure 14C the effect of dimension and hierarchy is analyzed. Interestingly, the small single-level array (solid squares) exhibits the highest adhesion performance for all preloads compared to the larger single-level structures (solid triangles) and the hierarchical design (solid circles). However, in particular for higher preloads, the adhesion of the hierarchical array increases greatly and finally reaches the same adhesion as the small single-level array. Unfortunately, the properties at higher preloads could not be investigated due to the relatively low spring constant of the cantilever (6.9 N m^{-1}). The lower adhesion performance of the hierarchical array up to a certain preload, presumably results from its smaller area fraction. For small preloads, the contact area is dominated by this area fraction. In this case less area fraction, therefore, leads to less adhesion. However, at higher preloads the larger pillars of the hierarchical structures can bend and so the compliance increases which results in increasing contact area and therefore higher adhesion values.

However, since the JKR-area during testing is predicted to be between 700 nm and 820 nm in diameter, the adhesion of the larger single-level pillars is curious at first sight. The area on top of the considered pillars is larger than the calculated JKR-area during testing. Hence, the measured adhesion of the larger single-level pillars should be comparable to the adhesion of the flat reference sample. The reason for the increased adhesion is revealed by the analysis of the adhesion maps shown in Figure 15C: the adhesion on top of the large single-level structures is very low and comparable to the adhesion measured on the flat reference sample, but is high at the structure edges. Since the used force measurement technique is not capable of distinguishing between different interactions we attribute the high adhesion at the structure edges mainly to mechanical interlocking. The diameter of the used spherical probe was small enough to sink into the gaps between the pillars. Consequently, a negative frictional force is detected during

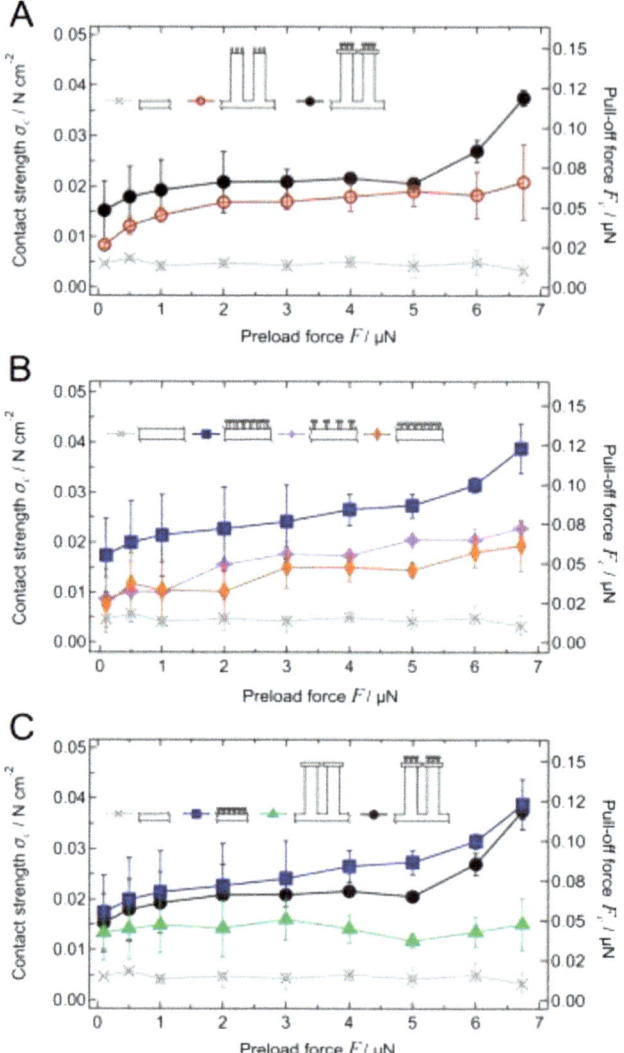

Figure 14 *The strong dependence of the adhesion on the actual design of the gecko-mimicking structures is revealed by the shown contact strength versus preload plots. Open symbols correspond to unstructured pillars while solid symbols refer to mushroom-shaped tips. A) The positive impact of the tip shape on the adhesion values is shown for the two folded hierarchical structures with (solid circles) and without (open circles) mushroom-shaped tips. B) The influence of density and aspect ratio is analyzed in this graph. The decrease of density (-20 %; solid stars) and aspect ratio (-10 %; solid diamonds) causes a significant reduction in adhesion by 40 % for a lower density and 50 % for a reduced aspect ratio. C) In this plot, the effect of the hierarchical design is investigated. The small single-level structure (solid squares) exhibits consistently the highest adhesion, closely followed by the hierarchical design (solid circles). The difference in the adhesion of those two arrays is probably induced by the larger area fraction of the small single-level structure. However, at higher preloads the hierarchical structures become more and more compliant whereby their lower area fraction is compensated.*

retraction due to clamping effects which is not to be confused with the perpendicular adhesion force originating from van-der-Waals forces. Figure 15 additionally shows the modification of the adhesion by adding the small single-level structures on top of the large single-level structures. This configuration leads to uniformly distributed adhesion all-over the array.

Typically, the adhesion of geckos is anisotropic, meaning the pull-off force necessary to detach a seta strongly depends on the direction of the preload. Preloading in both the perpendicular and parallel direction results in an over 10 times higher pull-off force of a single seta compared to preloading a seta just in the perpendicular direction.[3] Accordingly, geckos tend to pull their feet inwards toward their body causing a parallel preload of the setae. The resulting micron-scale displacement is necessary to align the setae, maximize contact area with the substrate and engage adhesion.[73] Autumn et al.[74] further demonstrated significant adhesive friction when isolated setae arrays were dragged along their natural curvature. In contrast, when dragged against their natural curvature much less friction exhibited and the adhesives pads are easily peeled from the surface.

The work of Autumn and coworkers inspired groups all over the world to mimic the directional adhesion in order to implement strong adhesion and easy release in synthetic gecko-like adhesives.[75-77] Various designs have been fabricated in order to switch adhesion. Angled fibers with angled tip endings made of soft polyurethane exhibited significant adhesion when loaded in one direction, and self-releasing behaviour when loaded in the opposite shear direction. Further, Murphy et al. demonstrated that the adhesion can be controlled by varying the shear displacement before loading in normal direction.[18] Shear-induced unidirectional adhesion has also been shown for stooped nanohairs and vertically aligned nanotubes with angled tips.[78,16] Reddy et al.[79] demonstrated a fiber array made of a shape-memory polymer. Thermally induced, the authors showed switchable adhesion by tilting the fibers. Wedge-shaped designs and angled fibers benefit from their reduced stiffness and enhanced compliance.[17,8] However, the resulting pull-off force for angled fibers is less due to a rotational moment that is induced additionally.[8] Hence, in order to exceed vertical fibers in terms of adhesion, angled fibers have to compensate the decreased pull-off force by an increased contact area. However, Reddy et al.[79] and Aksak et al.[8] measured less adhesion for angled fibers compared to vertical fibers with each were preloaded perpendicularly only.

By using 3D direct laser writing, we also fabricated tilted posts with mushroom-shaped tips in order to reveal the various options given by this fabrication technology (see Figure 16A-C). Subsequently, the tilted posts were examined with an AFM and our previously described method for measuring the adhesive forces. Since our testing protocol excludes lateral movement of the cantilever, the structures were preloaded in perpendicular direction only. The adhesion vs. preload curve displayed in Figure 16D compares the influence of tilting on the adhesion for small single-level structures and two-fold hierarchical structures as well. In agreement with the work of Reddy et al.[79] and Aksak et al.[8], the adhesion of tilted structures is less when being preloaded in perpendicular direction only. Adjusting the testing protocol in order to allow parallel preloads and measuring friction is ongoing work.

Finally, we would like to emphasize the robustness of the structures written by 3D direct laser writing. They survived thousands of force vs. distance curves taken with the AFM. Comparing SEM images of the structures before and after the adhesion measurements, we could not find any damaged or worn pillars (see Figure 17).

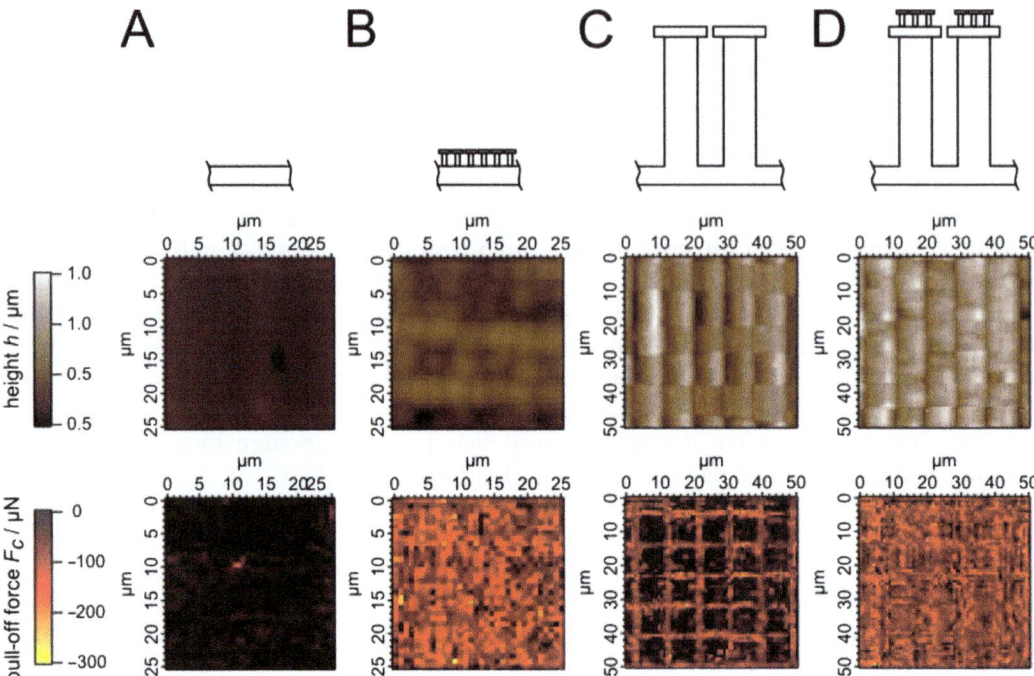

Figure 15 *Topography and adhesion measurements (preload: 6 µN) for the flat reference sample, the small and the larger single-level array and the hierarchical structure. The measurements were performed with an AFM using colloid tips with a diameter of 20 µm. A) The topography image of the reference sample shows its flatness in the nanometer range (root-mean-squared roughness measured with the colloid tip $R_q = 35$ nm) while the adhesion map highlights its poor adhesion. B) Due to the large radius of the colloid tip the topography of the small single level structures is not properly imaged. The adhesion, however, is greatly enhanced compared to the flat reference sample. C) The contour of the mushroom-shaped tips of the larger single-level array can be identified in the topography image. The corresponding force map, however, reveals that the tip frequently sticks in the gap between the pillars. D) This effect is prevented for the two-folded hierarchical structure. Since the probe cannot stick between the larger pillars, the hierarchical structure leads to uniformly distributed high adhesion.*

3   CONCLUSION

In summary, we proposed a method for the design of robust dry adhesives and presented a systematic design study and adhesion analysis of gecko-mimicking structures, which are close to the inspiring example of the gecko in dimensions and elastic modulus. With 3D direct laser writing it is straightforward to design, fabricate, and test arbitrary nano- and micrometer scale structures with or without hierarchy in a very flexible way. The interpretation of the presented results was supported by adhesion maps obtained with colloid AFM tips. The positive impact of mushroom-shaped tips could be demonstrated for stiff materials with lateral dimensions in the nanometer range.

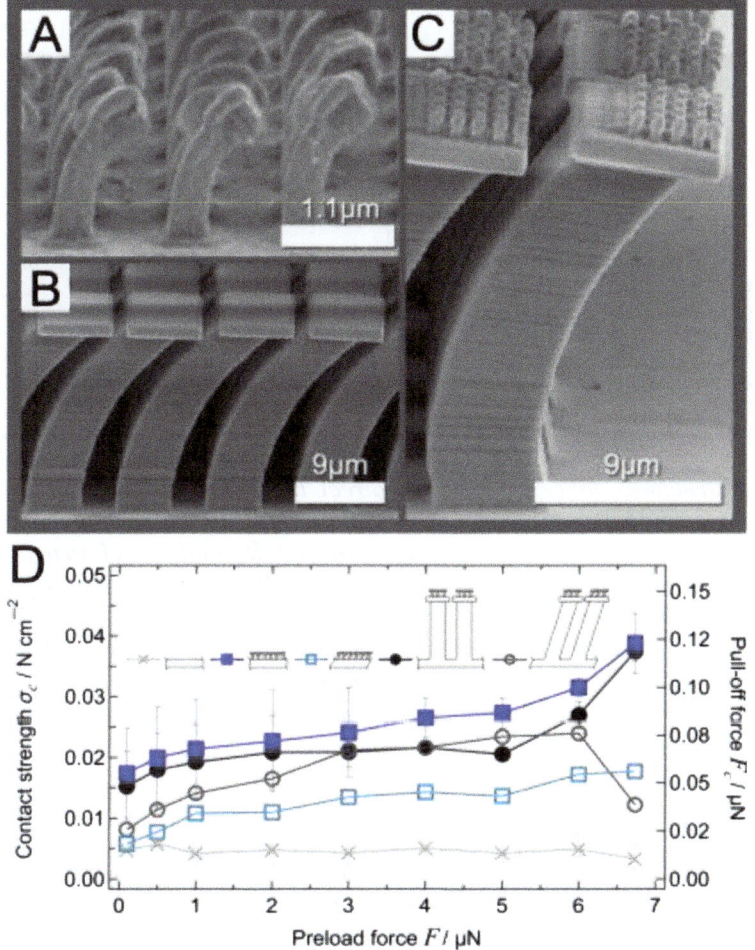

Figure 16 *A) Array 1, B) array 4 and C) array 7 were fabricated additionally with tilted posts. D) Compared to the straight arrays (solid symbols), an adhesion improvement could not be observed for the arrays with tilted pillars (open symbols).*

In addition, it was shown that the hierarchical structure of dry adhesives positively affects adhesion for appropriate preloads. This result supports the long-standing hypothesis that adhesion to natural rough surfaces requires a hierarchical design to ensure intimate contact and therefore a high overall amount of van-der-Waals forces.

In addition, 3D direct laser writing offers the possibility to fabricate three-dimensional templates, so that arbitrary but demoldable soft structures can be easily manufactured by casting. Therefore, we are convinced that the reported process will prospectively loom large in the fabrication of functional surfaces.

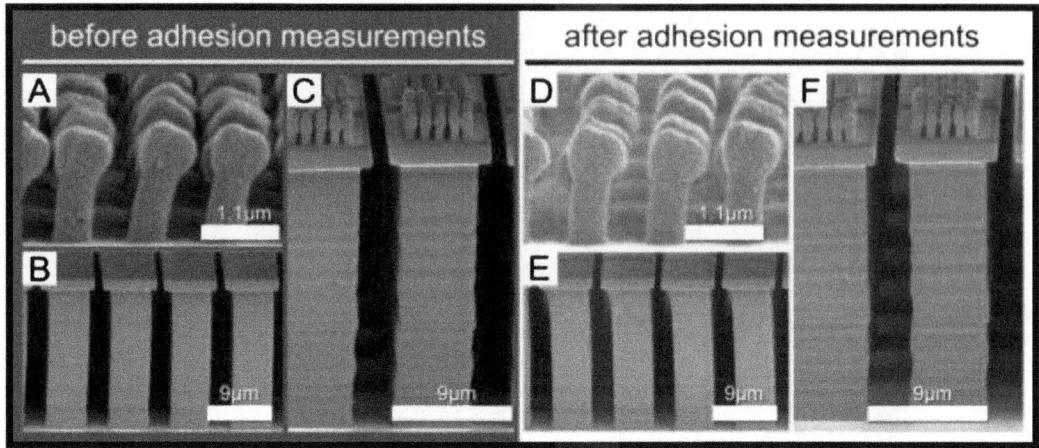

Figure 17 *The structures fabricated by 3D direct laser writing are very durable and showed no observable damage after thousands of adhesion measurements. A) SEM image of array 1, which contains small single level structures with a width of 500 nm and mushroom-shaped tips. B) The SEM image shows array 4, which contains mushroom-shaped pillars of 5 μm width. C) Array 7 consists of array 1 placed on top of array 4. The SEM images shown in A), B), and C) were taken before performing the adhesion measurements. D), E) and F) show the same arrays after the adhesion measurements.*

Acknowledgements

We thank Paul Abaffy for capturing the SEM images and Birgit and Reiner Kaup for learning more about Geckos. Furthermore, it is a pleasure to acknowledge fruitful discussions with Tobias Meier, Farid Oulhadj, Volker Saile and Juerg Leuthold. This work was partly carried out with the support of the Karlsruhe Nano Micro Facility (KNMF, www.kit.edu/knmf), a Helmholtz Research Infrastructure at Karlsruhe Institute of Technology (KIT, www.kit.edu).

References

1   R. Ruibal, V. Ernst, *J. Morphol.,* 1965, **117**, 271.
2   U. Hiller,*Z. Morph. Tiere,* 1968, **62**, 307.
3   K. Autumn, Y.A. Liang, S.T. Hsieh, W. Zesch, W.P. Chan, T.W. Kenny, R. Fearing, R.J. Full. *Nature,* 2000, **405**, 681.
4   P.Y. Hsu, L. Ge, X. Li, A.Y. Stark, C. Wesdemiotis, P.H. Niewiarowski, A. Dhinojwala, *J. R. Soc. Interface,* 2012, **9**, 657.
5   K. Autumn, C. Majidi, R.E. Groff, A. Dittmore, R. Fearing, *J. Exp. Biol.,* 2006, **209**, 3558.
6   K. Autumn, *Am. Sci.,* 2006, **March-April 2006**, 123.
7   K. Autumn, M. Sitti, Y.A. Liang, A.M. Peattie, W.R. Hansen, S. Sponberg, T.W. Kenny, R. Fearing, J.N. Israelachvili, R.J. Full, *Proc. Natl. Acad. Sci. USA,* 2002, **99**, 12252.
8   B. Aksak, M.P. Murphy, M. Sitti, *Langmuir,* 2007, **23**, 3322.
9   C. l. I. A. E. del Campo, A. Greiner, *Adv. Mater.,* 2007, **19**, 1973.

10 L. Ge, S. Sethi, L. Ci, P.M. Ajayan, A. Dhinojwala, *Proc. Natl. Acad. Sci. USA,* 2007, **104**, 10792.

11 S. Gorb, M. Varenberg, A. Peressadko, J. Tuma, *J. R. Soc. Interface,* 2007, **4**, 271.

12 C. Greiner, E. Arzt, A. del Campo, *Adv. Mater.,* 2009, **21**, 479.

13 H.E. Jeong, S.H. Lee, P. Kim, K.Y. Suh, *Nano Lett.,* 2006, **6**, 7 1508.

14 J. Lee, R.S. Fearing, *Langmuir,* 2008, **24**, 10587.

15 M.T. Northen, C. Greiner, E. Arzt, K.L. Turner. *Adv. Mater.,* 2008, **20**, 3905.

16 L. Qu, L. Dai, M. Stone, Z. Xia, Z.L. Wang, *Science,* 2008, **322**, 238.

17 A. Asbeck, S. Dastoor, A. Parness, L. Fullerton, N. Esparza, D. Soto, B. Heyneman, M. Cutkosky, *ICRA '09. IEEE International Conference on Robotics and Automation,* 2009, 2675–2680.

18 M.P. Murphy, B. Aksak, M. Sitti, *Small,* 2009, **5**, 170.

19 H. Ko, Z. Zhang, J.C. Ho, K. Takei, R. Kapadia, Y.-L. Chueh, W. Cao, B.A. Cruden, A. Javey, *Small,* 2010, **6**, 22.

20 Y. Zhang, C.-T.Lin, S. Yang, *Small,* 2010, **6**, 768.

21 M.K. Kwak, H.E. Jeong, W.G. Bae, H.-S.Jung, K.Y. Suh, *Small,* 2011, **7**, 2266.

22 Copyright Eric Isselée, 2012. Used under license from Shutterstock.com.

23 J. Batal, *Campus,* 2006, **01**,21.

24 A. del Campo, C. Greiner, E. Arzt, *Langmuir,* 2007, **23**, 10235.

25 M.P. Murphy, S. Kim, M. Sitti, *ACS Appl. Mater. Interfaces,* 2009, **1**, 849.

26 R. Spolenak, S. Gorb, H. Gao, E. Arzt, *Proc. R. Soc. Lond. A Math.,* 2005, **461**, 305.

27 R. Spolenak, S. Gorb, E. Arzt, *Acta Biomater.,* 2005, **1**, 5.

28 B. Bhushan, *J. Adh. Sci. Technol.,* 2007, **21**, 1213.

29 B.N.J. Persson, *J. Chem. Phys.,* 2003, **118**, 7614.

30 H. Yao, H. Gao, *J. Mech. Phys. Solids,* 2006, **54**, 1120.

31 M. Sitti, R.S. Fearing, *J. Adh. Sci. Technol.,* 2003, **17**, 1055.

32 C.-Y. Hui, A. Jagota, L. Shen, A. Rajan, N. Glassmaker, T. Tang, *J. Adh. Sci. Technol.,* 2007, **21**, 1259.

33 N.J. Glassmaker, A. Jagota, C.-Y.Hui, J. Kim, *J. R. Soc. Interface,* 2004, **1**, 1 23.

34 C.Y. Hui, A. Jagota, Y.Y. Lin, E.J. Kramer, *Langmuir,* 2002, **18**, 1394.

35 A. Jagota, S.J. Bennison, *Integr. Comp. Biol.,* 2002, **42**, 1140.

36 N.J. Glassmaker, A. Jagota, C.-Y.Hui, J. Kim, *J. R. Soc. Interface,* 2004, **1**, 35.

37 H. Gao, X. Wang, H. Yao, S. Gorb, E. Arzt, *Mech. Mater.,* 2005, **37**, 275.

38 N.J. Glassmaker, A. Jagota, C.-Y.Hui, *Acta Biomater.,* 2005, **1**, 367.

39 T. Tang, C.-Y.Hui, N.J. Glassmaker, *J. R. Soc. Interface,* 2005, **2**, 505.

40 H. Yao. *Ph.D. thesis,* 2006, Universität Stuttgart.

41 T.W. Kim, B. Bhushan, *J. Vac. Sci. Technol. A,* 2007, **25**, 1003.

42 M. Kamperman, E. Kroner, A. del Campo, R.M. McMeeking, E. Arzt, *Adv. Eng. Mater.,* 2010, **12**, 335.

43 C. Hui, Y. Lin, J. Baney, A. Jagota, *J. Adh. Sci. Technol.,* 2000, **14**, 1297.

44 B. Schubert, C. Majidi, R. Groff, S. Baek, B. Bush, R. Maboudian, R. Fearing, *J. Adh. Sci. Technol.,* 2007, **21**, 1297.

45 S. Liu, P. Zhang, X. Cheng, R. Malik, Z. Tang, *Adv. Sci. Lett.,* 2011, **4**, 1546.

46 K.L. Johnson, K.Kendall, A.D. Roberts, *Proc. R. Soc. Lond. A Math.,* 1971, **324**, 301.

47 E. Arzt, S. Gorb, R. Spolenak, *Proc. Natl. Acad. Sci. USA,* 2003, **100**, 10603.

48 D.G. Ellwood, *Mechanical Metallurgy,* 1961, New York, McGraw-Hill.

49 H.G. Hahn, *Bruchmechanik,* 1976, Teubner Studienbücher: Mechanik.

50 T.S.D. Gross, *Bruchmechanik,* 2006, Springer, Berlin.

51 M. Röhrig, M. Thiel, M. Worgull, H. Hölscher, *Small,* 2012, **8**, 3009.

52 M. Deubel, G. von Freymann, M. Wegener, S. Pereira, K. Busch, C.M. Soukoulis, *Nature Mater.,* 2004, **3**, 444.

53 A.M. Peattie, C. Majidi, A. Corder, R.J. Full, *J. R. Soc. Interface,*2007, **4**, 1071.

54 C. Greiner, A. del Campo, E. Arzt, *Langmuir,* 2007, **23**, 3495.

55 S.N. Gorb, M. Varenberg, *J. Adh. Sci. Technol.,* 2007, **21**, 1175.

56 A. Jagota, C.-Y. Hui, *Mater. Sci. Eng. R.,* 2011, **72**, 253.

57 N. Anscombe, *Nature Photon.,2010*, **4**, 22.

58 W. Oliver, G. Pharr, *J. Mater. Res.,* 2004, **19**, 3.

59 T. Kustandi, V. Samper, D. Yi, W. Ng, P. Neuzil, W. Sun, *Adv. Funct. Mater.,* 2007, **17**, 2211.

60 A.Y.Y. Ho, H. Gao, Y.C. Lam, I. Rodriguez, *Adv. Funct. Mater.,* 2008, **18**, 2057.

61 A.Y.Y. Ho, L.P. Yeo, Y.C. Lam, I. Rodriguez, *ACS Nano,* 2011, **5**, 1897.

62 H.-J. Butt, B. Cappella, M. Kappl, *Surf. Sci. Rep.,* 2005, **59**, 1.

63 L.H. Mak, M. Knoll, D. Weiner, A. Gorschluter, A. Schirmeisen, H. Fuchs, *Rev. Sci. Instrum.,* 2006, **77**, 046104.

64 E. Kroner, D.R. Paretkar, R.M. McMeeking, E. Arzt, *J. Adhesion,* 2011, **87**, 447.

65 G. Castellanos, E. Arzt, M. Kamperman, *Langmuir,* 2011, **27**, 7752.

66 D. Tabor, *J. Colloid Interf. Sci.,* 1977, **58**, 2.

67 B. Derjaguin, V. Muller, Y. Toporov, *J. Colloid Interf. Sci.,* 1975, **53**, 314.

68 D. Maugis, *Contact, Adhesion and Rupture of Elastic Solids*, 2000,Springer-Verlag.

69 J.N. Israelachvili, *Intermolecular and Surface Forces*, 1998, Academic Press.

70 V.L. Popov, *Kontaktmechanik und Reibung*, 2009, Springer Berlin.

71 J.A. Greenwood, J.B.P. Williamson, *Proc. R. Soc. Lond. A Math.,* 1966, **295**, 300.

72 G. Carbone, E. Pierro, S.N. Gorb, *Soft Matter,* 2011, **7**, 5545.

73 K. Autumn, A.M. Peattie, *Integr. Comp. Biol.*, 2002,**42**, 1081.

74 K. Autumn, A. Dittmore, D. Santos, M. Spenko, M. Cutkosky, *J. Exp. Biol* , 2006, **209**, 3558.

75 D. Santos, M. Spenko, A. Parness, S. Kim, M. Cutkosky, *J. Adh. Sci. Technol.*, 2007, **21**, 1317.

76 J. Lee, C. Majidi, B. Schubert, R.S. Fearing, *J. R. Soc. Interface*, 2008, **5**, 835.

77 M.P. Murphy, B. Aksak, M. Sitti, *J. Adh. Sci. Technol.*, 2007, **21**, 1281.

78 T.-I. Kim, H.E. Jeong, K.Y. Suh, H.H. Lee, *Adv. Mater.,* 2009, **21**, 2276.

79 S. Reddy, E. Arzt, A. del Campo, *Adv. Mater.,* 2007, **19**, 3833.

UNDERWATER ADHESION OF MUSHROOM-SHAPED ADHESIVE MICROSTRUCTURE: AN AIR-ENTRAPMENT EFFECT

E. Kizilkan*, L. Heepe and S.N. Gorb

*Department of Functional Morphology and Biomechanics, Zoological Institute at the University of Kiel, Am Botanischen Garten 1–9, 24098 Kiel, Germany
*ekizilkan@zoologie.uni-kiel.de

1   INTRODUCTION

The research on adhesive properties of various biological systems e.g. attachment structures of geckos, spiders, flies, and beetles is the newly developed and actively growing field of adhesion science.[1,2] During the past decade many adhesive micro-structured systems have been produced mimicking these biological systems.[3] One of them is the mushroom-shaped adhesive microstructure (MSAMS) (Figure 1) inspired by the adhesive tarsal hairs of male beetles of the family Chrysomelidae. This particular industrial development demonstrated several important advantages compared to conventional pressure sensitive adhesives. It was previously shown that MSAMS exhibited about twice higher pull-off force measured on smooth substrates, if compared to a smooth control sample made from the same material.[4] Moreover, it retained its adhesive capability over thousands of attachment cycles and after being contaminated an initial adhesion capability could be recovered by washing it in light soap solution in water.[5] The superiority of MSAMS has been previously demonstrated in a comparative study that revealed that adhesion strength of MSAMS on smooth surfaces is close to that of the gecko toe.[6] The enhanced adhesion of MSAMS was attributed to the combination of the intermolecular van der Waals forces and its particular crack trapping geometry.[4,5]

Under water the problem of displacing water from the adhesive interface makes adhesion complicated due to weakened chemical bonds.[7] However, MSAMS also showed unexpectedly high underwater adhesion which was even higher than that measured in air.[8,9] As it was previously indicated by considering Hamaker constants of MSAMS-glass contact in air and underwater, contribution of van der Waals forces under water to the adhesion is reduced by about 86%.[10] In order to explain the enhanced adhesion performance of MSAMS, a passive suction effect that is potentially possible under each individual microstructure was proposed.[8]

However, our recent observations on living beetles[11] have demonstrated that the air entrapment effect might be responsible for adhesion enhancement of MSAMS under water. The aim of the present study was an experimental proof of this hypothesis. For this purpose, we compared pull-off forces of MSAMS under water with entrapped air with those completely wetted by water and with those in dry conditions in the air.

Figure 1 *Scanning electron microscope (SEM) images of mushroom-shaped adhesive microstructure (MSAMS) inspired by the adhesive pad structures of male beetles from the family Chrysomelidae.*

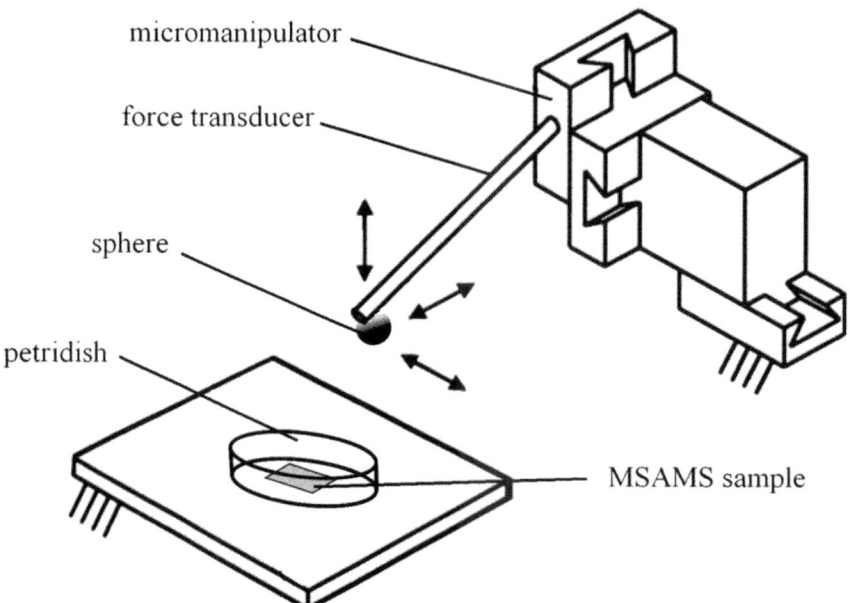

Figure 2 *Experimental set-up. MSAMS sample is attached to the bottom of the Petri dish. The contact was established with the sapphire sphere attached to the tensomeric force transducer. Force measurements were carried out with the force transducer attached to a motorized micromanipulator (adapted from 12).*

2 MATERIALS AND METHODS

The MSAMS material used in this study was made out of polyvinylsiloxane (PVS). The sample surface was covered by hexagonally distributed individual MSAMSs of about 100 μm height terminated by a thin contact plate covering about 50% of the apparent contact area.[5]

Pull-off forces were measured using a custom-made setup, similar to that reported in Varenberg and Gorb (2008)[12], consisting of a motorized micromanipulator (DC 3001R, World Precision Instruments Inc., Sarasota, Florida, USA) and a fixed tensometric force transducer FORT-100 (World Precision Instruments Inc., Sarasota, Florida, USA) with a 3 mm sapphire sphere attached to it. A MSAMS sample was placed in a Petri dish (Figure 2). After preloading of the MSAMS sample with the sphere in the load range of about 0.1 – 40 mN, pull-off forces were determined while withdrawing the sphere from the contact at a velocity of 200 μm/s. In order to measure the effect of entrapped air on underwater adhesion of MSAMS, the following test procedure was performed. First, pull-off forces were determined on samples in the dry state. Then, water was carefully added into the Petri dish. At the sites, where air was entrapped between single MSAMS, pull-off forces were measured again. Finally, the entrapped air was removed by gently rubbing the sample with a spatula, and the pull-off forces of the fully wetted sample were measured again. Eight individual samples were measured on at least 30 different sites in each condition.

In order to verify different wetting states of MSAMS in contact to a substrate, an additional experiment was performed. A MSAMS sample was brought into contact to a smooth cover slip and the contact interface was observed with an inverted microscope Observer.A1 (Carl Zeiss MicroImaging GmbH, Göttingen, Germany) operating in an epi-illumination mode. The contact interface was first visualized for the dry state and subsequently in underwater condition with and without entrapped air.

The static contact angle of water on the surface of MSAMS and sapphire sphere were measured by using high-speed optical contact angle measuring device OCAH 200 (Data Physics Instruments GmbH, Filderstadt, Germany). The sessile drop method (drop volume 1 μl) with circle fitting was used to calculate the water contact angle (water density=1000 kg.m^{-3}). For each surface, the contact angle measurements were repeated at six different sites.

3 RESULTS AND DISCUSSION

Figure 3 shows the contact interface between the MSAMS sample and the smooth glass surface in the dry state (Figure 3a) and under water (Figure 3b,c). Under water, the MSAMS sample may clearly have two different wetting states. The dark areas represent the real contact zone of the terminal lips of the MSAMS with the cover slip. The MSAMS submerged in water reveals its ability to retain air between its pillars (Figure 3b), even in contact with a solid substrate. This air retaining state of the MSAMS under water represents an example of the Cassie wetting state.[13] Without air entrapment, the MSAMS is in the Wenzel wetting state.[14]

The pull-off force measurements revealed different adhesive properties of MSAMS samples in different states (Figure 4). Pull-off forces were averaged for the preloads in the ranges of 0-10 mN, 11-20 mN, 21-30 mN, and 31-40 mN. The measured pull-off forces on the MSAMS under water with air entrapment were on average about 5.5 times higher if compared to those measured in the fully wetted state of the MSAMS. Furthermore, the air entrapped state of the MSAMS exhibited about 2.5 times higher pull-off forces, if compared to the dry state. Additionally, a clear preload dependence of pull-off forces has been observed, except for the fully wetted state. In both dry and air-entrapped states, the higher the applied preloads, the higher were the pull-off forces. This dependence can be

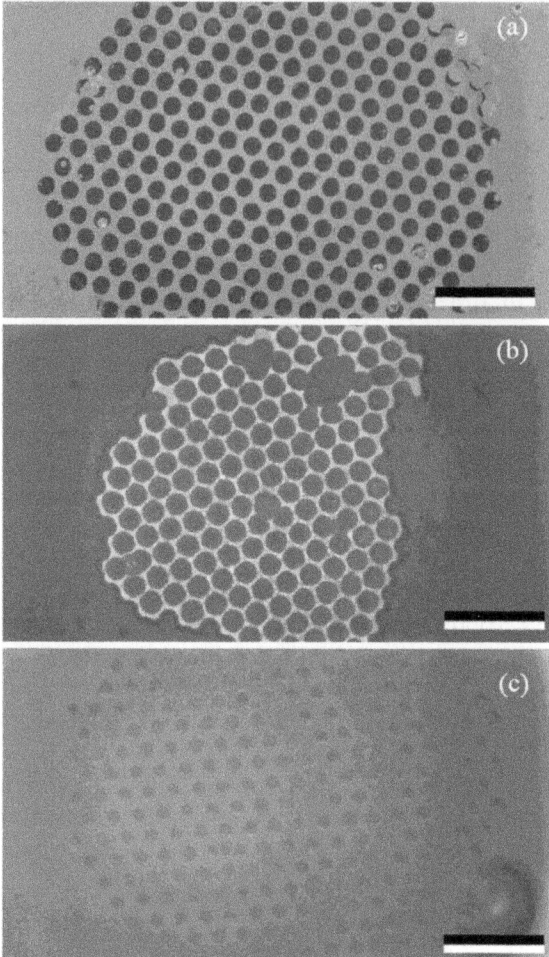

Figure 3.   *Contact interface of the MSAMS in contact with a smooth glass cover slip at different wetting states. Reflection contrast microscopy images: (a) dry state, (b) in water with entrapped air, and (c) in water fully wetted. Dark areas correspond to real contact zones. Scale bar is 250 μm.*

well described by the Schargott-Popov-Gorb (SPG) model of contact mechanics of fibrillar attachment devices.[15] By enlarging the indentation radius of the sphere in contact with the fibrillar surface, the contact area increases with an increasing preload, resulting in higher pull-off forces. During the underwater experiment with the entrapped air, Cassie-Wenzel state transformation did not occur and the air layer was stable for the entire range of preloads applied.

In order to understand the experimental results obtained, diverse physical mechanisms of adhesion of the MSAMS were considered. In general, adhesion ability of the MSAMS was attributed to rely mainly on intermolecular van der Waals forces and particular contact geometry rather than capillary forces.[5,8,12,16] It can be assumed here that intermolecular van der Waals forces at the PVS-glass contact should be weakened by about 86%, when the contact is submerged in water.[8,10] Higher pull-off forces under water were previously hypothesized to be caused by the individual MSAMSs acting as passive suction devices:

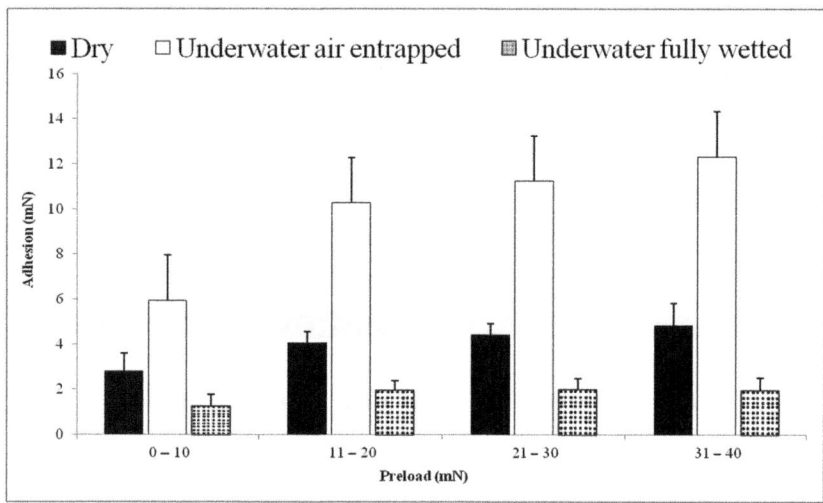

Figure 4 *Pull-off forces of the MSAMS measured at the different wetting states at different preloads.*

the negative pressure under each individual MSAMS can cause cavitation effect due to microfibre stalk tension leading to the transformation of liquid water into vapor. Such a transformation may give rise to a suction effect.[8]

Based on the present results we cannot exclude such underwater suction effect for individual MSAMSs. Nevertheless, since the pull-off forces in the fully wetted state were much lower than those in the air entrapped state and even lower than in dry conditions, we hypothesize that the amount of underwater adhesion of samples with an entrapped air cannot be solely explained by the suction effect mentioned above. In our experiment, we used MSAMS made of PVS which is inherently hydrophobic with a water contact angle of ~120°, and a sapphire sphere with a water contact angle of ~93°.[17,18] In such a condition, when both MSAMS and the substrate are hydrophobic, we assume MSAMS in the region of entrapped air to be virtually in dry condition due to de-wetting of the contact interface (Figure 5a). This hypothesis is supported by recent observations of the underwater adhesion in the terrestrial leaf beetle *Gastrophysa viridula* having similar microstructures on its adhesive feet as in our experiment[11] and as in some other artificial microstructures.[11-19] In the experiments with beetles as well as with artificial microstructures with an air entrapment, pull-off forces under water increased with increasing degree of hydrophobicity of the substrate. Nevertheless, solely de-wetting does not explain the about 2.5 times higher pull-off forces in the air entrapped state compared to the dry state in our experiments. When air is entrapped and contact areas are de-wetted, the pull-off forces would be expected to be similar to the dry state, but not significantly higher. Therefore, enhanced underwater adhesion should be explained by an additional mechanism.

We assume that the intermolecular van der Waals forces are superimposed with another suction effect caused by the air entrapment between individual MSAMSs (Figure 5a, b). During pull-off, the volume of entrapped air V_0 with a pressure p_0 increases due to elastic deformations of individual MSAMSs. As a consequence, the pressure p in the entrapped air spaces decreases, inducing a pressure difference to the surrounding pressure and thus results in a global suction effect caused by the air bubble deformation. This suction force is predicted to be the highest just before detachment, when both the volume

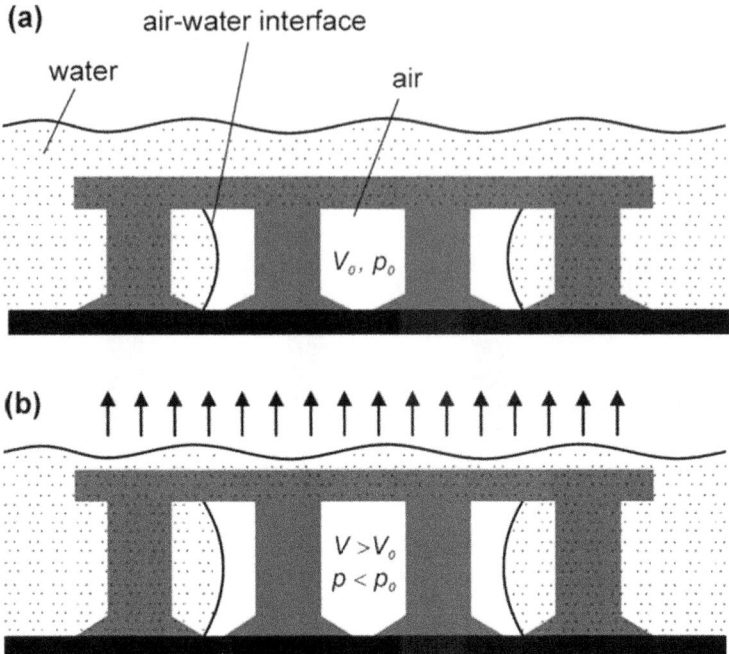

Figure 5. *Schematic of the air entrapment effect on the MSAMS, when submerged in water at equilibrium (a) and during pull-off (b). In the contact with a hydrophobic substrate, air entrapping areas on MSAMS effect in de-wetting of the substrate. V_0, initial volume of entrapped air. p_0, initial pressure in the entrapped air volume. V and p, volume and pressure during pull-off.*

of the air entrapping space and pressure difference, p-p_0, are greatest. A further influence of hydrostatic pressure must not be considered, since at a maximum depth of only 1 cm under water, the hydrostatic pressure contributes only about 0.1 % additionally to the atmospheric pressure and could therefore be neglected.

4   CONCLUSION

In the present work, we performed pull off force measurements on the MSAMS at different wetting states, in order to shed light on the mechanism of enhanced underwater adhesion of MSAMS. It has been demonstrated that under water the MSAMS is able to entrap air inducing stronger underwater adhesion than in both dry and fully wetted conditions. We believe that the air entrapment effect on MSAMS adhesion may lead to the design of novel switchable adhesive tapes for medical, marine, and many other kinds of underwater applications.

Acknowledgements

This project was funded by the German Research Foundation (DFG) under the grant scheme SFB 677-C10.

References

1 S.N. Gorb, *Attachment Devices of Insect Cuticle*, Kluwer Academic Publishers, 2001.
2 S.N. Gorb, *Handbook of Adhesion Technology*, Springer, Berlin, Heidelberg 2011, 1410.
3 D. Sameoto,, and C. Menon, *Smart Mater. Struct.*, 2010, **19**, 103001.
4 L. Heepe, A.E. Kovalev, M. Varenberg and S.N. Gorb, *Theor. Appl. Mech. Lett.*, 2012, **2**, 14008.
5 S.N. Gorb, M. Varenberg, A. Peressadko and J. Tuma, J., *J. R. Soc. Interface*, 2007, **4**, 271.
6 A. del Campo, C. Greiner and E. Arzt, *Langmuir*, 2007, **23**, 10235.
7 A.M. Smith and J.A. Callow, *Biological Adhesives*. Springer, Berlin, 2006.
8 M. Varenberg and S.N. Gorb, *J. R. Soc. Interface*, 2008, **5**, 383.
9 P. Glass, H. Chung, N.R. Washburn and M. Sitti, *Langmuir*, 2010, 26, 17357.
10 J. Israelachvili, in *Intermolecular and Surface Forces*, NY: Academic Press, New York, 1992, 176.
11 N. Hosoda and S.N. Gorb, S. N., *Proc. R. Soc. B.*, 2012, **279**, 4236.
12 M. Varenberg and S.N. Gorb , *J. R. Soc. Interface*, 2008, **5**, 785.
13 A.B.D. Cassie and S. Baxter, *Trans. Faraday Soc.*, 1944, **40**, 546.
14 R.N. Wenzel, R. N., Ind. Eng. Chem. 1936, **28**, 988.
15 M. Schargott, V.L. Popov and S.N. Gorb, S.N. *J. Theor. Biol.*, 2006, **243**, 48-53.
16 S.N. Gorb and M. Varenberg, *J. Adhesion Sci. Technol.*, 2007, **21**, 1175.
17 R. van Noort, *Introduction to Dental Materials*, Mosby, Spain, 1994.
18 R.G. Craig, W.J. O'Brien and J.M. Powers, *Dental Materials. Properties and Manipulation*, 6th edition., Mosby, St Louis, 1996
19 D.M. Drotlef, S. Lukas, M. Kappl, W.J.P. Barnes, H.J. Butt and A. del Campo, *Adv. Funct. Mater.*,2012, DOI: 10.1002/adfm.201202024

BIO-INSPIRED DUPLEX ATTACHMENT PAD WITH ASYMMETRIC ADHESION

J.Y. Chung*[1,2] and M.K. Chaudhury[1]

[1] Lehigh University, Department of Chemical Engineering, 111 Research Drive, Bethlehem, 18015, Pennsylvania, USA
[2] Harvard University, School of Engineering and Applied Sciences, 60 Oxford Street, Cambridge, 02138, Massachusetts, USA
*jchung@seas.harvard.edu

1 INTRODUCTION

The roles of the segmentation and discontinuity of adherends in the overall strength of adhesive joints have been investigated for a long time in the field of materials science.[1] Early studies by Kendall[2] and more recent ones by us[3–5] and others[6–9] have demonstrated that a strip of rubber, which has either discontinuities in thickness, stiffness or work of adhesion, or singularities arising from a segmented surface, exhibits higher peeling strength than a smooth and homogeneous rubber. Various factors, such as the crack growth retardation due to the sharp gradient of strain energy, crack path deflection and contact splitting caused by surface structuring, are responsible for enhancing adhesion strength in such cases. Similarly, it has been shown that self-organized systems, including crazing in thermoplastics[10] and fibrillation in pressure sensitive adhesives,[11] can lead to the toughening of interface by distributing the external force to smaller domains.

The principles underlying the 'flaw tolerance' of interface that is crucial to obtaining high fracture toughness in human-engineered adhesive systems are currently being interpreted in the context of biological adhesive systems.[12–14] New insights have been gleaned from nature in the past decade,[15–19] which have fueled the rapid development of bio-mimetic and bio-inspired adhesives in which micro/nano structures are built on synthetic materials.[20–24] However, unlike biological adhesive systems, synthetic structured adhesives (for example, those based on a gecko-seta-mimetic fibrillar interface[24]) often fail to perform optimally due to the limited flexibility, which may not allow intimate contact with rough surfaces. Enhanced contact compliance of soft and/or slender fibrillar structures may compensate for surface roughness, but these structures tend to undergo lateral collapse or buckling.[25,26] To address these shortcomings, researchers have begun to identify the important role of the upper-level element of biological hierarchical architecture (*i.e.*, contact tip) and have proposed a variety of bio-inspired strategies, including film-terminated[27] and mushroom-shaped[28] tips built on fibrillar structures.

Herein, we focus on a relatively less-explored aspect—the role of the lower-level element of hierarchical architecture inherent in biological adhesive systems (*i.e.*, supporting matrix). The feet of many animals and insects have attachment pads that are composed of a smooth or textured skin supported by a spongy-like matrix.[29,30] Inspired by these biological footpads, we investigate the adhesive properties of a bilayer system[31] consisting of a highly compliant support terminated with a thin elastic film and, in particular, the effect of the elastic compliance of the support on contact and adhesion behavior. One potential benefit of such a bilayer system is that the high compliance of the support is expected to allow for the top film to achieve good conformal contact even with uneven surfaces. Another benefit is that the soft support may render the top film a high strain tolerance. As will be seen below, the bilayer system studied in the present work shows interesting adhesive properties, exhibiting strong pull-off adhesion to rough and curved surfaces, but sufficiently weak adhesion to permit easy peeling. This combination of imparting high adhesion in a normal pull-off mode but poor adhesion in a peel mode motivates us to further explore the design of a duplex attachment pad that yields asymmetric adhesion.

2 MATERIALS AND METHODS

2.1 Polydimethylsiloxane (PDMS) Block

5 mm-thick blocks of PDMS were prepared using a 10:1 (by weight) mixture of oligomer and curing agent (Sylgard 184, Dow Corning Co.). A thoroughly hand-stirred mixture was poured onto a glass Petri dish, degassed for 30 min in a vacuum, and then cured at 120 °C for 1 h in a pre-heated convection oven. The resulting cross-linked PDMS was then cut using sharp steel punches into circular blocks of 10 mm in diameter (for pull-off adhesion measurements) and rectangular blocks of 25 mm in width and 50 mm in length (for peel adhesion measurements). One side of each PDMS block was plasma-oxidized using the Harrick plasma cleaner and then bonded to a cleaned and plasma-oxidized rigid glass substrate that was attached to a rigid frame. The frame, in turn, can be mounted on the load cell of the adhesion testing system. The Young's modulus of the cross-linked PDMS was measured to be 1.5 MPa, and its Poisson's ratio was assumed to be 0.5.

2.2 PDMS/Polyurethane (PU) Bilayer

200 μm-thick films of PDMS were prepared following a previously established protocol.[32] In brief, two spacers with a thickness of 200 μm were placed at each end of a glass microscope slide (50 mm × 75 mm × 1 mm; Corning Science Products) to achieve a uniform film thickness. The glass slide was pre-coated with a self-assembled monolayer (SAM) of hexadecyltrichloro silane (United Chemicals Technologies Inc.) to serve as an easy release layer. A well-stirred mixture of PDMS oligomer and curing agent (at a ratio of 10:1 by weight) was poured on the SAM-coated glass slide between two spacers and allowed enough time (about 2 h) to spread evenly over the slide. The mixture was then cured at 70 °C in a relatively short time period (about 10–15 min) to yield slightly cross-linked networks of PDMS. The highly compliant support used in this study was a commercially available open-cell PU foam (5 mm in thickness). The Young's modulus of the PU foam was 0.1 MPa, which was measured by an indentation method. The Poisson's ratio of the PU foam was assumed to be 0.3. The PU foam was carefully brought onto contact with the surface of the partially cured PDMS thin film and subsequently heated at elevated temperature (120°C) for 1 h to produce completely cross-linked networks of PDMS. This procedure prevents the PDMS mixture from penetrating deeply into the open cell PU foam, but a slight penetration is permissible since it enhances the interfacial bonding between film and foam. The glass slide coated with the SAM was then carefully peeled off, leaving only the flat, thin PDMS film attached to one side of the PU foam.

Finally, the resulting PU foam-supported thin PDMS film was then cut into cylindrical (10 mm in diameter; for pull-off adhesion measurements) and rectangular shapes (25 mm in width and 50 mm in length; for peel adhesion measurements) using sharp steel punches. The other side of the PU foam had an adhesive, which was brought into a rigid frame and gently pressed to secure good interfacial bonding between PDMS/PU bilayer and rigid frame. The frame that holds the PDMS/PU bilayer, in turn, can be attached to the load cell of the adhesion testing system.

2.3 Rough and Curved Surface

Glass microscope slides (50 mm × 75 mm × 1 mm; Corning Science Products) were used for their microscopically smooth surface (root-mean-square roughness less than 4 nm)[33] and also used to produce rough surfaces. All glass slides were chemically cleaned with toluene and acetone, and then dried under pure nitrogen flow. A series of rough surfaces with varying microscale roughness were obtained by blasting the cleaned glass slides with 50 μm alumina grit for 10 s. The grit was fluidized in pressurized air of three different pressures (0.20 MPa, 0.41 MPa, and 0.62 MPa), which yielded three different average surface roughnesses (R_a = 13 μm, 18 μm, and 32 μm), respectively. The roughness of the grit-blasted slides was measured by scanning surface profiles with the optical profilometer (Stil Micromeasure, Micro Photonics). The average surface roughness was quantified as the root-mean-square roughness by measuring the arithmetical mean deviation of the assessed altitude profile. For curved surfaces, optically smooth glass hemispherical lenses with three different radii of curvature (R = 1.7 cm, 6 cm, and 12 cm) were used in this study. The curved glass surfaces were chemically cleaned with toluene and acetone, and then dried under pure nitrogen flow before all measurements.

2.4 Pull-off Experiment

For pull-off adhesion measurements, we used a specifically designed, axisymmetric pull-off apparatus[32] that allows precise measurements of force and displacement. The measurements of adhesion (normal pulling) force between a flat-ended cylindrical sample (PDMS block or PDMS/PU bilayer) and a rigid substrate (having smooth, rough, or curved surfaces) were conducted as follows. The contact between the sample and the substrate was maintained for 30 min. The sample was then retracted from the contacting substrate at a constant displacement rate of 5 μm/s by using a displacement control system. During debonding, a load cell allowed the simultaneous acquisition of voltage data, which were converted to force data using the computer-assisted data-acquisition system (model PCI-DAS6035; Measurement Computing Co.) *via* LabView software. The nominal pulling stress ($\sigma = F/\pi a^2$, a being the radius of the cylindrical sample) and nominal strain ($\varepsilon = \delta/a$) were determined from the measured normal pulling force (F) and the vertical displacement of the sample relative to the position at zero force displacement (δ), respectively. Note that the adhesion strength between a sample and a rough surface and between a sample and a curved surface was found to depend strongly on the initial preload; thus, the same preload (200 g) was applied to all measurements.

2.5 Peel Experiment

For peel adhesion measurements, we used the displacement-controlled, peel adhesion apparatus described previously.[5] The model cantilever used in this study was a microscope glass slide (flexural rigidity, D, = 6 Nm; Corning Science Products) with the following dimensions: width = 25 mm, length = 75 mm, and thickness = 1 mm. In particular, a SAM-coated glass slide was used in peel adhesion measurements instead of the uncoated one used in pull-off adhesion measurements, because the latter can produce enough adhesion force to exceed the limitation of load cell capacity used. We note that this was not the case for the pull-off experiment, in which a much smaller initial contact area was used (about 16 times smaller than that for the peel experiment). A part of the SAM-coated glass cantilever (25 mm in width and 50 mm in length; thus, the initial contact area, $A = 1.25 \times 10^{-3}$ m^2) was initially brought in complete contact with the surface of a rectangular sample of PDMS block or PDMS/PU bilayer. The initial cantilever arm length (the distance between the point where the vertical displacement is applied to deform the cantilever and the closest contact edge) was fixed at $l_0 = 13$ mm. After waiting for 30 min, one end of the contacting cantilever (where force was measured) was then lifted at a constant rate of displacement (5 µm/s) using a nanomotion controller system coupled with a computer-assisted data acquisition module. This system allows us to control the vertical displacement (Δ) of the cantilever end as well as to simultaneously measure the peeling force (F). The resultant force-displacement curve was recorded and further analyzed by using LabView software.

3 RESULTS AND DISCUSSION

3.1 Pull-off Adhesion

Our model bilayer system consists of a 200 µm-thick cross-linked film of PDMS supported by a 5 mm-thick PU foam (PDMS/PU bilayer). We examine its adhesive properties through pull-off and peel tests and compare them with those of a 5 mm-thick PDMS block. Figure 1 shows the results of pull-off adhesion measurements for both sets of samples. In our experiment with a cylindrical PDMS block (5 mm in radius) against a smooth glass substrate, the pull-off stress (σ^*), or the maximum adhesion strength, is measured to be about 16 kPa. In the case where a thick deformable disk is pulled from a rigid substrate (or a rigid disk from a thick deformable substrate), the crack initiates at the edge of the contact and propagates toward the center, and the theoretical pull-off stress is given by:[32,34,35]

$$\sigma^* = \left[\frac{8WE}{\pi a (1 - \nu^2)} \right]^{1/2} \tag{1}$$

where a is the apparent radius of contact, E and ν are the Young's modulus and Poisson's ratio of the deformable material, respectively, and W (= $\gamma_1 + \gamma_2 - \gamma_{12}$) is the work of adhesion. Here γ_1 and γ_2 are the surface energies of the contacting bodies, and γ_{12} is the interface energy of the two surfaces. Based on Equation 1, we estimate the work of adhesion between PDMS ($E = 1.5$ MPa, $\nu = 0.5$) and glass to be 250 mJ/m^2. This value is larger than the work of adhesion expected for dispersion interaction (44 mJ/m^2), suggesting that there is a specific interaction between PDMS and glass.[36,37]

Given the structural difference between the samples, it is striking that there is no discernible difference in their pull-off stress (see Figure 1b). However, if we compare their stress-strain curves, we find a distinct difference. In contrast to the linear behavior of the PDMS block, the PDMS/PU bilayer exhibits a nonlinear stress-strain behavior with a

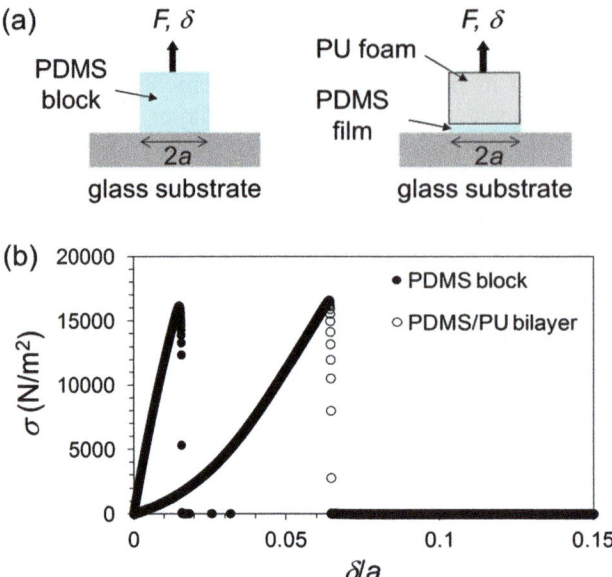

Figure 1 *(a) Schematic of the punch pull-off experiment, in which a flat-ended cylindrical sample (PDMS block and PDMS/PU bilayer) with a radius of a was initially placed in contact with a microscopically smooth glass substrate and then detached from the substrate at a constant speed. The normal pulling force (F) and the vertical displacement (δ) were simultaneously recorded during debonding. (b) Representative plots of nominal stress (σ = F/πa²) vs. nominal strain (ε = δ/a) for two sets of samples.*

tendency to stiffen at larger strains. We postulate that such strain-stiffening can be induced by the confinement of the stiff top film adhered on the soft support, which causes the crack to initiate and grow from the interior rather than the edge of the contact, thus rendering a high strain tolerance with improved adhesion. In an extreme case, if we suppose that the thin PDMS film attached on the thick PU foam behaves like a fully confined film, the pull-off stress is given by:[32,35]

$$\sigma^* = \left(\frac{3.3WE}{h}\right)^{1/2} \qquad (2)$$

where h is the film thickness. According to Equation 2 with $W = 250$ mJ/m², $E = 1.5$ MPa, and $h = 200$ μm, the pull-off stress is about 80 kPa. On the other extreme, if we consider the deformability of the PDMS/PU bilayer as controlled solely by that of the PU foam, the pull-off stress is determined to be about 4 kPa by Equation 1 with the mechanical properties of the PU foam ($E = 0.1$ MPa, $v = 0.3$). The measured pull-off stress (17 kPa) is in between the above two extremes; thus, the confinement effect somehow comes into play. An exploration of the complex coupling between the top and bottom layers is needed to fully understand this effect, as has been demonstrated earlier,[38,39] but here we treat our bilayer as being, to a rough first approximation, a single entity with an effective modulus. From the stress–stain plot for the PDMS/PU bilayer shown in Figure 1b, we find that the effective tensile modulus for PDMS/PU is initially about 0.12 MPa, which is close to the Young's modulus of the PU foam (0.1 MPa), but increases gradually with increasing nominal strain and reaches to about 0.55 MPa beyond a strain of about 0.05. Here, the values of the effective modulus at a given strain level ($\varepsilon = \varepsilon_c$) were estimated using the following equation: $E(\varepsilon_c) = \pi(1-v^2)/2 \cdot \partial\sigma/\partial\varepsilon\big|_{\varepsilon_c}$, where v is the Poisson's ratio (assumed to be 0.3) and $\varepsilon \,(= \delta/a)$ is the nominal strain. Using Equation 1, together with $E = 0.55$ MPa, we estimate the pull-off stress for the PDMS/PU bilayer to be about 10 kPa, slightly smaller than the measured value of 17 kPa.

The discrepancy between the measured and theoretical values is likely due to the rough approximation applied, but there could be several other factors to be considered. For instance, it is plausible that in this kind of bilayer structure, a shear stress could develop at the interface, and thus, some energy might be consumed in interfacial friction. Enhancement of adhesion of a thin PDMS film by friction has been reported previously.[40] It is also plausible that the adhesion energy of the bilayer could, in fact, be enhanced somewhat because it can conform to surface roughness much better due to its improved contact compliance compared to PDMS alone. We will return to this point shortly. However, we make at this point an important observation that the stored elastic energy (per unit area) for the deformed PDMS/PU bilayer ($U = 2.03$ J/m^2) is much larger than that of the PDMS block ($U = 0.67$ J/m^2). We note that these energy values were determined by calculating the area under the stress-strain curves ($U = \int \sigma d\delta$) in Figure 1b. It would be possible to achieve this result with a material of low modulus; however, the stress to failure will decrease with decreasing modulus according to $\sigma \sim E^{1/2}$. Thus, a bilayer structure exhibiting nonlinear stiffening with increasing deformation ensures a high stress and high fracture energy.

3.2 Effects of Surface Roughness and Surface Curvature

We now turn to the issues of the surface roughness as well as the radius curvature of the substrate surface and examine how each affects contact and adhesion behavior. Figure 2 shows the configuration and the results of pull-off adhesion measurements for both sets of samples against rough glass substrates having three different surface roughnesses (R_a). In general, it is evident that the pull-off stress for both samples decreases as roughness increases. However, roughness-induced decrease in pull-off stress is more pronounced with the PDMS block as compared to the PDMS/PU bilayer. Intuitively, these results suggest that at a given load, a softer film is expected to make a better contact with a rough surface.

From purely dimensional arguments, we can reasonably assume that the fractional area of contact will depend on W, E, and R_a as $(W/ER_a)^\alpha$ (α being an arbitrary positive constant); *i.e.*, softer the film and lower the roughness, greater is the real area of contact. Because of the high compliance of the PU support, we should expect the PDMS/PU bilayer to show a better contact with a rough surface. In fact, it is not just that the contact area is greater for each asperity. Given the distribution of surface asperities of different sizes, many of the asperities would not make contact with the PDMS block, but a greater number of asperities would make contact with the PDMS/PU bilayer. A similar conclusion has recently been drawn from both theoretical analysis[41] and experimental results[42] of layered materials in contact with rough surfaces.

There could be another possibility as to why the pull-off stress could increase with compliance (inverse of mechanical stiffness), which, to the best of our knowledge, has not

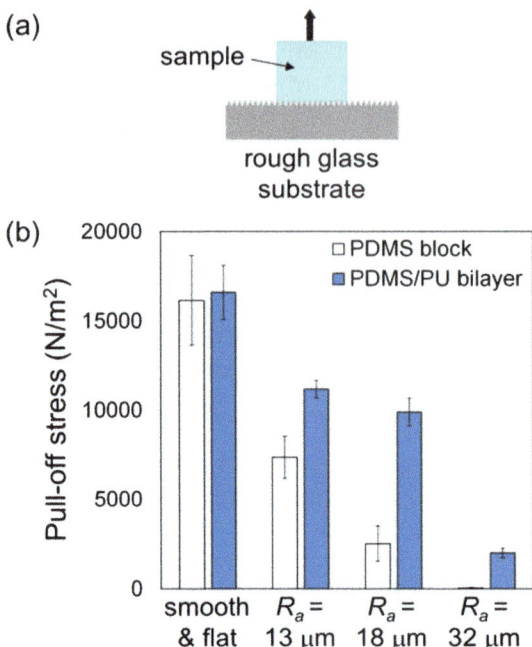

Figure 2 *(a) Schematic of a flat-ended cylindrical sample in contact with a rigid substrate with a grit-blasted rough surface. (b) The measured pull-off stresses plotted as a function of surface roughness (R_a). The error bars represent one standard deviation of the data, which is taken as the experimental uncertainty of the measurement.*

been pointed out in the literature. Here we briefly review the various geometric configurations available in the literature and discuss the dependence of the elastic modulus on the pull-off force in each configuration. For the classical Johnson-Kendall-Roberts (JKR) geometry,[43] including a sphere on a flat substrate, two spheres in contact and two perpendicular cylinders, the pull-off force is independent of modulus. For a cylinder on a flat substrate or two parallel cylinders,[44–46] the pull-off force increases with the one-third power of modulus. For a flat-ended cylinder on a flat substrate,[32,34] the pull-off force increases with the square root of modulus. For a cone on a flat substrate,[47] the pull-off force increases inversely with modulus.

By examining various geometric combinations, we find that the only case that the pull-off force decreases with modulus is for a cone-shape tip detaching from a flat rubber, where there is no geometric length scale to describe stress amplification. According to Maugis and Barquins,[47] the pull-off force for this case is given by $F^* = 81W^2/2\pi E\tan^3\beta$, where β is the conical tip angle. Since randomly rough surfaces prepared by grit-blasting in this study have surface roughness features with sharp tips, the detachment force of such an asperity is reasonably well described by the above equation. Thus, softer the film, stronger is the adhesion force per asperity. However, as the average spacing of these asperities is on the order of the root-mean-square roughness, the nominal pull-off stress would scale as $\sigma \sim W^2/ER_a^2$. Note that this relation is particularly true for periodic structures. With this being the case, it is likely that the pull-off stress would increase with the compliance, but decrease with the surface roughness.

Figure 3 shows the configuration and the results of pull-off adhesion measurements obtained with curved glass substrates having three different radii of curvature. It appears that the PDMS block loses adhesion much faster with decreasing the radius of curvature than the PDMS/PU bilayer does. This result is quite similar to the observed roughness effect in the sense that the easy deformability of PDMS/PU provides better contact on a curved surface as compared to the PDMS block having a higher bending modulus. In this case, however, there could also be another subtle effect due to friction. In a geometry such as in Figure 3a, the crack is expected to propagate by the growth of cavitation bubbles and a low-angle peeling is likely to occur at/near the crack tip with the possibility of the enhancement of fracture energy by kinetic friction.[40] However, this particular problem should be studied in great detail in future work.

3.3 Peel Adhesion

Having discussed the adhesion behavior of the PDMS block and the PDMS/PU bilayer in the pull-off geometry, we now direct our focus towards their adhesion behavior in the peel geometry, which is assessed by using a single cantilever beam test geometry (Figure 4a). Figure 4b shows the peeling force per unit width (F/w) as a function of displacement (Δ) for both sets of samples. Perhaps the most salient feature in Figure 4b is that the maximum peel-off force, or the crack initiation force, of the PDMS/PU bilayer is considerably smaller than that of the PDMS block. This result is quite intriguing, since it shows a trend

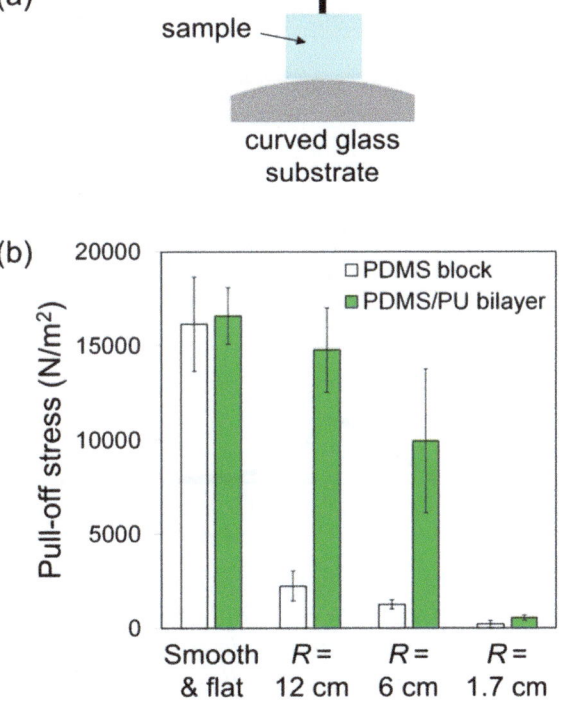

Figure 3 *(a) Schematic of a flat-ended cylindrical sample in contact with a rigid substrate whose contacting surface is curved. (b) The measured pull-off stresses plotted as a function of the radius of curvature (R). The error bars represent one standard deviation of the data, which is taken as the experimental uncertainty of the measurement.*

markedly different from the one for the pull-off behavior given in Figure 1b. We will discuss this seemingly paradoxical finding below, but before proceeding, we make an important observation. As seen in Figure 4b, the peeling behavior of both the samples does not follow the classical behavior of the peeling of a flexible cantilever beam from a rigid substrate or a rigid beam from a flexible substrate. Instead, it resembles the pull-off behavior in the sense that once the force reaches a critical value for crack initiation, a catastrophic failure occurs. In order to understand this situation more clearly, we make a rough estimate for the stress decay length (d) of a glass cantilever peeling from a sample. In this case, $d \approx (4Dh/E)^{1/4}$,[48] where D is the flexural rigidity of the cantilever and h is the sample thickness ($D = 6$ Nm and $h = 5$ mm in this study). Using the values of E ranging from 0.1 MPa (for PU foam) to 1.5 MPa (for PDMS block), we find that 17 mm $\leq d \leq 22$ mm, which is about one-half of the sample dimension (50 mm in length in this study). Since almost the half contact zone is stressed, an external torque may be generated at the interface. Because of this reason, we expect that the peeling of a glass cantilever from a sample more closely resembles the 'shearing' or 'tilting' off of a rigid stud from the edge of the interface between a stud and a sample.[49,50]

We can derive an expression for the strain energy release rate in this configuration, following a procedure similar to that described earlier.[49,50] One of the two necessary

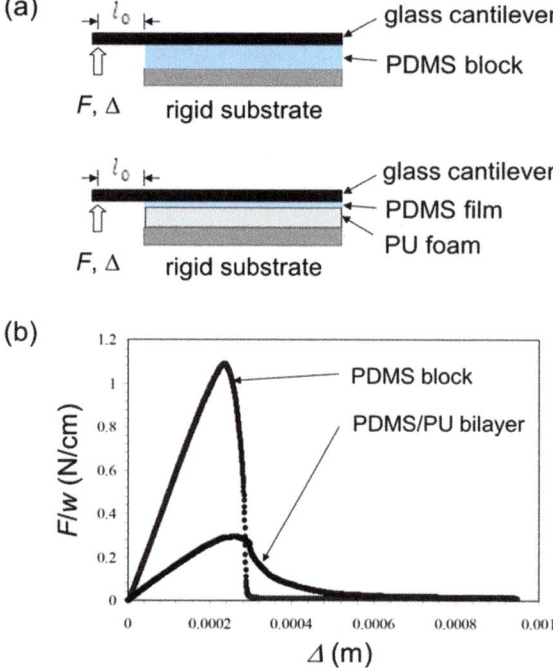

Figure 4 *(a) Schematic of the displacement-controlled peel experiment, in which a silanized glass cantilever with an initial arm length of l_0 was initially brought into contact with a rectangular sample (PDMS block and PDMS/PU bilayer) and then its one end was lifted vertically at a constant speed. The peeling force (F) and the vertical displacement (Δ) were simultaneously recorded during peeling. (b) Representative plots of force per unit width of the cantilever (F/w) vs. displacement (Δ) for two sets of samples.*

equations is the balance between the external and internal torques and the other is the total energy of the system, from both of which the strain energy release rate can be determined. After completing this exercise, one can find that the peel-off force per unit width (F^*/w) is expressed as:

$$\frac{F^*}{w} \approx \frac{l^2}{l_0 + l}\left(\frac{WE}{h}\right)^{1/2} \tag{3}$$

where l_0 is the initial arm length and l is the lateral length of the contact ($l_0 = 13$ mm and $l = 50$ mm in this study). For a silanized glass peeling from the PDMS block ($W = 44$ mJ/m^2 and $E = 1.5$ MPa), we find that $F^*/w \approx 1.4$ N/cm from Equation 3. The value is in close agreement with the measured value of 1.1 N/cm, indicating that Equation 3 performs well. We note that the slight difference between the measured and calculated values might be due to the fact that we neglected all the prefactors in the derivation of Equation 3.

Figure 4b shows that F^*/w for the PDMS/PU bilayer is about 0.3 N/cm. Based on this value and Equation 3, we estimate its effective modulus to be about 0.07 MPa. Very interestingly, the estimated value is similar to the effective modulus of the PDMS/PU bilayer at low strain levels (0.12 MPa). This result suggests that the effect of confinement in the peeling case is nonexistent or not as strong as in the pull-off case. In the absence of confinement or at its very low degrees, edge crack propagation is likely to dominate, thus not allowing the bilayer to withstand large strains.

3.4 Asymmetric Adhesion

The results presented above clearly show the influence of the backing layer and illustrate the importance of the actual loading conditions in the adhesion strength of layered systems. We interpret that the observed difference between the pull-off and peel-off forces of the PDMS/PU bilayer arises from its varying modulus for a specific experimental geometry as it exhibits nonlinear stress-strain response. On the one hand, as the peeling behavior for the PDMS/PU layer is controlled by its lower low-strain modulus, the peel-off force is much smaller than that of the PDMS block. On the other hand, as the pull-off behavior of the PDMS/PU layer is controlled by its higher high-strain modulus, there is very little difference in the pull-off force between the two samples.

The combination of imparting high adhesion in a pull-off mode but easy release in a peel mode with the bilayer system inspires us to further explore the design of a duplex attachment pad possessing a compliance gradient along the peeling direction. In particular, we consider a duplex attachment pad that consists of two geometrically identical wedge-shaped blocks with different elastic modulus. As illustrated in Figure 5a, these blocks are stacked together in such a way that the diagonal side of the wedge of one block of a higher modulus (PDMS elastomer) is supported by that of the other block with a low modulus (PU foam). Thus, the total thickness of the duplex attachment pad remains constant, but there exists a gradient in effective compliance along a long axial direction due to the opposing thickness gradients of the two blocks having different moduli. We examine the adhesion behavior of the resulting duplex attachment pad using a simple experimental setting in which a weighted object is initially placed on the pad surface and then a rigid substrate strongly adhered to the back side of the attachment pad is grasped with one hand and lifted up (Figure 5a), rotated counterclockwise (Figure 5b), or rotated clockwise (Figure 5c).

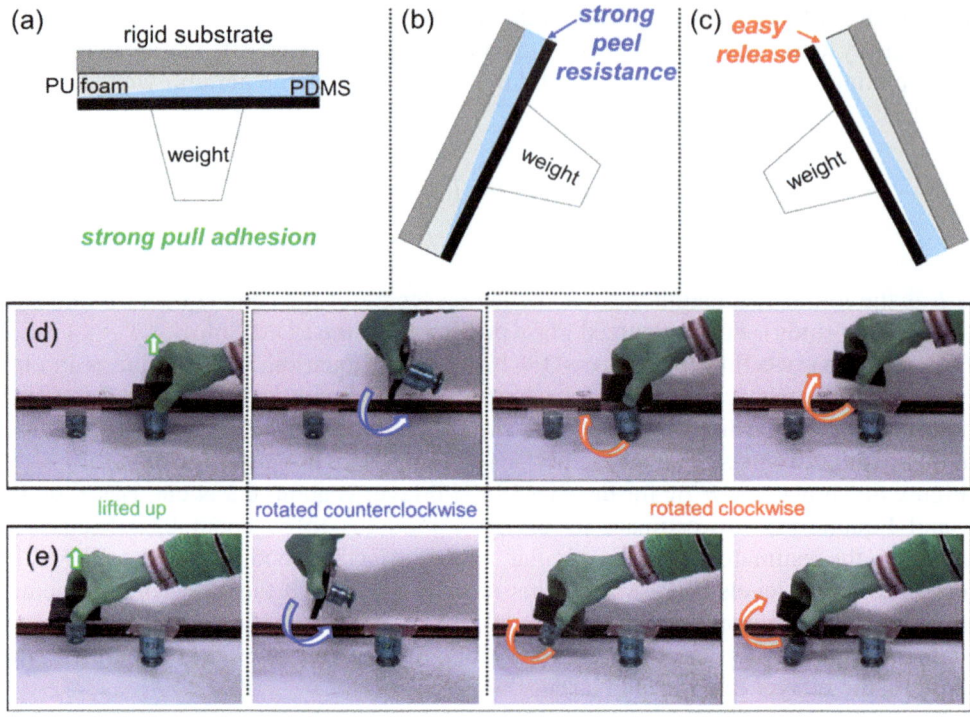

Figure 5 *(a–c) Schematic illustrations of the three possible adhesion behaviors of the duplex attachment pad in which the thicknesses of both PDMS elastomer and PU foam vary linearly from opposite ends in an axial direction (50 mm in length): (a) strong adhesion in a normal pull-off mode, (b) strong resistance when peeled from the thick to the thin PDMS end, and (c) easy peeling from the thin to the thick PDMS end. (d,e) Two series of snapshots revealing the asymmetric adhesion behavior of the duplex attachment pad contacting with a heavy object of mass M: (d) M = 500 g and (e) M = 200 g.*

Experiments involving the duplex attachment pad (Figures 5d and e) reveal a couple of interesting results and some potential paths for future research. First, the duplex attachment pad shows strong adhesion in a normal pull-off mode (see the first column of Figures 5d and e). This result can be explained based on the results presented in Figure 1b that both the PDMS block and PDMS/PU bilayer are able to withstand a similar but considerable load in the normal direction. Second, the duplex attachment pad shows asymmetric adhesion behavior, revealing strong peeling resistance when peeled from the thick to the thin end of the PDMS (*i.e.*, when rotated counterclockwise; see the second column of Figures 5d and e), but easy release in the opposite direction (*i.e.*, when rotated clockwise; see the third and fourth columns of Figures 5d and e). These results can be interpreted based on the results presented in Figure 4. In the peeling situation, the thicker PDMS region possessing a lower level of mechanical compliance exhibits a greater peeling resistance than the thinner PDMS region that has a relatively higher compliance. In other words, the separation of interface can be either accelerated or retarded, depending on the direction of the compliance gradient.

4 CONCLUDING REMARKS

In this study, we illustrate that a bilayer system consisting of a thin elastic film supported on a highly compliant support exhibits interesting adhesive properties comprising non-linear stress-strain behavior, resulting in strong pull-off adhesion to rough and curved surfaces, but sufficiently weak adhesion to permit easy peeling. Inspired by these results, we consider the design of a duplex attachment pad, in which the thicknesses of both PDMS elastomer and PU foam vary gradually from opposite ends. Such a duplex system reveals asymmetric adhesion, showing strong peeling resistance when peeled from the thick to the thin end of the PDMS, but weak peeling resistance in the opposite direction.

Our foam-based design strategy has significant potential in various applications,[51,52] such as reusable wall-climbing robots, wafer handling, and self-cleaning adhesives, where directional adhesion capability is of primary importance. We anticipate that the differential adhesive properties can be further improved through optimization and/or by incorporating micro/nano structures often involved in biological and bio-mimetic adhesives (for example, those based on an anisotropically patterned or tilted fibrillar interface[53-55]). Equally importantly, the use of a compliant support on which rigid fibrillar structures are built will also likely alleviate potential robustness problems often encountered when using soft fibrillar structures to enhance contact compliance. Finally, the nonlinear mechanics of an adhesive presents itself an interesting design principle even at micro-structural level. For example, if the fibrils in the gecko-mimetic tapes are deigned to exhibit non-linear stress-strain behavior, either exploiting its geometry or its material property, a high level of detachment stress and, accordingly a high level of fracture energy may be obtained from the collective behaviors of the fibrils.

Acknowledgement

This work was supported in part by the Office of Naval Research (ONR). We thank Dr. Jiong Liu for help in preparing the rough substrate surfaces.

References

1 J. E. Gordon, *The New Science of Strong Materials, or Why You Don't Fall through the Floor*, Princeton University Press, Princeton, 1976.
2 K. Kendall, *Proc. R. Soc. A*, 1975, **341**, 409.
3 M. K. Chaudhury, T. Weaver, C. Y. Hui and E. J. Kramer, *J. Appl. Phys.*, 1996, **80**, 30.
4 A. Ghatak, L. Mahadevan, J. Y. Chung, M. K. Chaudhury and V. Shenoy, *Proc. R. Soc. A*, 2004, **460**, 2725.
5 J. Y. Chung and M. K. Chaudhury, *J. R. Soc. Interface*, 2005, **2**, 55.
6 A. Ghatak, *Phys. Rev. E*, 2010, **81**, 021603.
7 S. Vajpayee, K. Khare, S. Yang, C. Y. Hui and A. Jagota, *Adv. Funct. Mater.*, 2011, **21**, 547.
8 S. Xia, L. Ponson, G. Ravichandran and K. Bhattacharya, *Phys. Rev. Lett.*, 2012, **108**, 196101.
9 A. Majumder, S. Mondal, A. K. Tiwari, A. Ghatak and A. Sharma, *Soft Matter*, 2012, **8**, 7228.
10 C. Creton, E. J. Kramer, C. Y. Hui and H. R. Brown, *Macromolecules*, 1992, **25**, 3075.
11 H. Lakrout, P. Sergot and C. Creton, *J. Adhes.*, 1999, **69**, 307.
12 K. Kendall, *Molecular Adhesion and Its Applications: The Sticky Universe*, Kluwer Academic/Plenum Publishers, New York, 2001.

13 H. J. Gao, B. H. Ji, I. L. Jager, E. Arzt and P. Fratzl, *Proc. Natl. Acad. Sci. USA*, 2003, **100**, 5597.
14 C. Y. Hui, N. J. Glassmaker, T. Tang and A. Jagota, *J. R. Soc. Interface*, 2004, **1**, 35.
15 K. Autumn, Y. A. Liang, S. T. Hsieh, W. Zesch, W. P. Chan, T. W. Kenny, R. Fearing and R. J. Full, *Nature*, 2000, **405**, 681.
16 E. Arzt, S. Gorb and R. Spolenak, *Proc. Natl. Acad. Sci. USA*, 2003, **100**, 10603.
17 W. Federle, *J. Exp. Biol.*, 2006, **209**, 2611.
18 W. Federle, W. J. P. Barnes, W. Baumgartner, P. Drechsler and J. M. Smith, *J. R. Soc. Interface*, 2006, **3**, 689.
19 S. N. Gorb, *Phil. Trans. R. Soc. A*, 2008, **366**, 1557.
20 M. Sitti and R. S. Fearing, *J. Adhes. Sci. Technol.*, 2003, **17**, 1055.
21 A. K. Geim, S. V. Dubonos, I. V. Grigorieva, K. S. Novoselov, A. A. Zhukov and S. Y. Shapoval, *Nature Mater.*, 2003, **2**, 461.
22 H. Lee, B. P. Lee and P. B. Messersmith, *Nature*, 2007, **448**, 338.
23 L. T. Qu, L. M. Dai, M. Stone, Z. H. Xia and Z. L. Wang, *Science*, 2008, **322**, 238.
24 N. J. Glassmaker, A. Jagota, C. Y. Hui and J. Kim, *J. R. Soc. Interface*, 2004, **1**, 23.
25 D. Chandra and S. Yang, *Acc. Chem. Res.*, 2010, **43**, 1080.
26 A. Jagota and C. Y. Hui, *Mater. Sci. Eng. R*, 2011, **72**, 253.
27 N. J. Glassmaker, A. Jagota, C. Y. Hui, W. L. Noderer and M. K. Chaudhury, *Proc. Natl. Acad. Sci. USA*, 2007, **104**, 10786.
28 G. Carbone, E. Pierro and S. N. Gorb, *Soft Matter*, 2011, **7**, 5545.
29 S. N. Gorb, *Attachment Devices of Insect Cuticle*, Kluwer Academic Publishers, Dordrecht, Netherlands, 2001.
30 A. Majumder, A. Ghatak and A. Sharma, *Science*, 2007, **318**, 258.
31 J. Y. Chung and M. K. Chaudhury, *Adhesive Behavior of Segmented Films*, Proceedings of the Annual Adhesion Society Meeting, Jacksonville, FL, 2006.
32 J. Y. Chung and M. K. Chaudhury, *J. Adhes.*, 2005, **81**, 1119.
33 M. K. Chaudhury and J. Y. Chung, *Langmuir*, 2007, **23**, 8061.
34 K. Kendall, *J. Phys. D: Appl. Phys.*, 1971, **4**, 1186.
35 J. Y. Chung, K. H. Kim, M. K. Chaudhury, J. Sarkar and A. Sharma, *Eur. Phys. J. E*, 2006, **20**, 47.
36 K. Vorvolakos and M. K. Chaudhury, *Langmuir*, 2003, **19**, 6778.
37 A. Ghatak, K. Vorvolakos, H. Q. She, D. L. Malotky and M. K. Chaudhury, *J. Phys. Chem. B*, 2000, **104**, 4018.
38 J. Yoon, C. Q. Ru and A. Mioduchowski, *J. Appl. Phys.*, 2005, **98**, 113503.
39 R. Mukherjee, R. Pangule, A. Sharma and G. Tomar, *Adv. Funct. Mater.*, 2007, **17**, 2356.
40 B. M. Z. Newby and M. K. Chaudhury, *Langmuir*, 1998, **14**, 4865.
41 B. N. J. Persson, *J. Phys.: Condens. Matter*, 2012, **24**, 095008.
42 D. Voigt, A. Karguth and G. Gorb, *Robotics and Autonomous Systems*, 2012, **60**, 1046.
43 K. L. Johnson, K. Kendall and A. D. Roberts, *Proc. R. Soc. A*, 1971, **324**, 301.
44 D. Maugis and M. Barquins, *J. Appl. Phys. D*, 1978, **11**, 1989.
45 J. A. Greenwood and K. L. Johnson, *Philos. Mag.*, 1981, **43**, 697.
46 M. Barquins, *J. Adhes.*, 1988, **26**, 1.
47 D. Maugis and M. Barquins, *J. de Phys. Lett.*, 1981, **42**, L95.
48 M. F. Kanninen, *Int. J. Fract.*, 1973, **9**, 83.
49 M. K. Chaudhury and K. H. Kim, *Eur. Phys. J. E*, 2007, **23**, 175.
50 M. K. Chaudhury and J. Y. Chung, *Langmuir*, 2007, **23**, 8061.
51 S. N. Gorb, M. Sinha, A. Peressadko, K. A. Daltorio and R. D. Quinn, *Bioinsp. Biomim.*, 2007, **2**, S117.

52 M. P. Murphy, C. Kute, Y. Menguc and M. Sitti, *Int. J. Robotics Res.*, 2011, **30**, 118.

53 K. Autumn, A. Dittmore, D. Santos, M. Spenko and M. Cutkosky, *J. Exp. Biol.*, 2006, **209**, 3569.

54 M. P. Murphy, B. Aksak and M. Sitti, *Small*, 2009, **5**, 170.

55 H. E. Jeong and K. Y. Suh, *Soft Matter*, 2012, **8**, 5375.

TARGETING SPECIFIC APPLICATIONS

WHAT CAN WE LEARN FROM THE OCTOPUS?

F. Tramacere*[1,2], L. Beccai[1] and B. Mazzolai[1]

[1]Center for Micro-BioRobotics@SSSA, Istituto Italiano di Tecnologia, Viale Rinaldo Piaggio 34, 56025 Pontedera, Italy
[2]BioRobotics Institute, Scuola Superiore Sant'Anna, Viale Rinaldo Piaggio 34, 56025 Pontedera, Italy
*francesca.tramacere@iit.it

1 INTRODUCTION

In robotics, reliable reversible attachment mechanisms can be used for both the locomotion of robots in unstructured environments and for object manipulation. The suction cup is one of the most commonly used reversible attachment mechanisms in these fields[1-4]. This device is used for a wide range of purposes because of its industrial-strength reliability, its excellent grip (up to 1 atm)[5], and its ease of use[6]. There is an enormous interest in developing these devices for application in many fields, spanning industrial manipulation[7] to object holders[8] to medical use[9-10].

Traditional suction methods generally involve the use of magnetic fields[11] or air pumps[12-13]. Both of these solutions present drawbacks: the former are only suitable for surfaces made of magnetic materials, while the latter are characterised by loud noise and large size. Some attempts have been made to overcome these limitations. A robot named Dexter, which creates a suction force by pressing passive suction cups to a wall using a motor, has been designed[14]. Another invention uses a vibrating suction method, whereby continuously pressed suction cups are maintained by a motor in high-frequency vibration[15].

However, currently, no solution can even remotely compare to the attachment mechanisms offered by nature, such as octopus suckers. The octopus sucker is an amazing example of flexibility, dexterity, and the ability to generate large forces on different substrates. Octopus suckers work in distinct ways and perform a remarkable variety of functions, such as in locomotion, anchoring the body of the octopus to the substratum while holding on to prey, manipulating small objects, etc[16]. Two devices have been inspired by this natural mechanism[17]. These devices work under dry conditions without an air pump and are actuated by a shape memory alloy (SMA). The first prototype uses a two-way shape memory effect extension titanium-nickel spring that mimics the piston structure of the stalked sucker; the second prototype is actuated by a one-way SMA actuator with a bias that has a basic structure of a stiff margin, a guiding element, a leader and an elastic element. Despite their innovative designs, these mechanisms are less efficient (corresponding to a load capacity of almost 2 N) than the natural attachment mechanism (which has a load capacity of tens of Newtons).

We have been inspired by the octopus sucker as a key biological attachment model. We are particularly interested in the octopus' ability to adhere to wet surfaces, which can

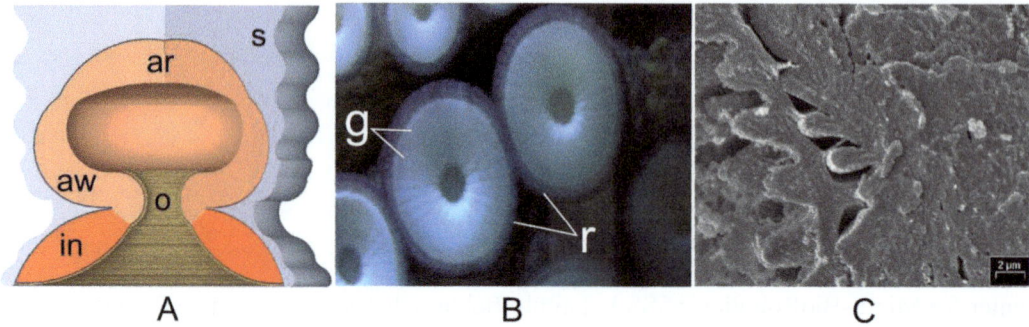

Figure 1   A) *Schematic of an octopus sucker: skin, s; acetabular roof, ar; acetabular wall, aw; orifice, o; and infundibulum, in. B) Photograph of octopus suckers: detail of radial grooves, g; and rim, r. C) Cryo-SEM image showing the renewal of an infundibular chitinous cuticle; the scale bar corresponds to 2 μm.*

prevent other types of adhesive mechanisms (such as certain glues), and in the hypothesised ability of this animal to remain attached to the substratum over an extended time period without any energy consumption[18-19]. For these reasons, we investigated the morphological and physiological features of this natural adhesive mechanism to formulate new design criteria for innovative bio-inspired adhesion devices. In this work, we present the first passive prototype to mimic the ability of octopus suckers to attach to different substrates and a design of an active suction cup inspired by the long-term attachment strategy of the octopus. Both the prototypes work underwater and exploit the incompressibility of water to obtain very strong adhesion without being large in size.

2 MORPHOLOGY AND PHYSIOLOGY OF THE OCTOPUS SUCKER

The octopus sucker is a muscular-hydrostat[20] that consists of two main parts: the acetabulum and the infundibulum. The acetabulum is a hollow spherical cup, and the infundibulum is a disk-like object that contacts the substrate during adhesion[18-20] (Figure 1A). The infundibular surface is rough with radial grooves and is encircled by a rim (Figure 1B). The external surface of the infundibulum is covered by a chitinous cuticle or sucker lining that is periodically shed and continuously renewed [16, 21-23] (Figure 1C).

The acetabulum and infundibulum are connected by an orifice, and both parts play an important role in the adhesion process.

The sucker musculature is arranged in a three-dimensional array consisting of three types of muscular fibres: meridional, circular and radial. The meridional fibres radiate outwards from the apex of the acetabular roof and extend down along the entire sucker (Figure 2A). The circular muscle bundles are oriented parallel to the opening surface of the infundibulum (Figure 2B) and the radial muscle fibres extend throughout the thickness of the sucker (Figure 2C).

Kier and Smith constructed an adhesion hypothesis[18-19] that the infundibulum, being dexterous and flexible, can be compliant with almost any surface by creating a seal during adhesion (Figure 3A); the complementary acetabulum functions as a pump, reducing the pressure inside the sucker cavity and thereby inducing suction (Figure 3B).

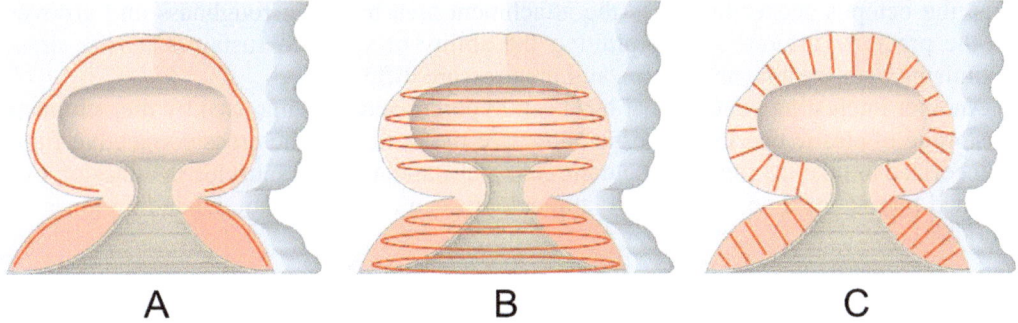

Figure 2 *Muscular arrangement in the octopus sucker: A) meridional fibres radiate outwards from the apex of the acetabular roof and extend down along the entire sucker; B) circular fibres are oriented parallel to the opening surface of the infundibulum; and C) radial fibres extend throughout the thickness of the sucker.*

This octopus adhesion mechanism is facilitated by the following: (i) the roughness of and the grooves in the infundibular surface and (ii) the incompressibility of water. (i) The roughness and grooves of the infundibular surface enhance adhesion at the substrate interface by creating an interconnected water-filled network of spaces that transmits the sub-ambient pressure induced by the acetabulum to the entire infundibulum[18]. The attachment area (A) is thus maximised. (ii) The ability of water to sustain a tensile stress enables the acetabulum to induce a pressure difference beyond that of a vacuum, as is the case with an air-filled sucker. The pressure difference (ΔP) is consequently maximised.

The attachment force (F) is equal to the pressure difference (ΔP) multiplied by the attachment area (A), so that the force of attachment (F) (1) increases as the two variables (ΔP and A) increase.

$$F = \Delta P * A \qquad (1)$$

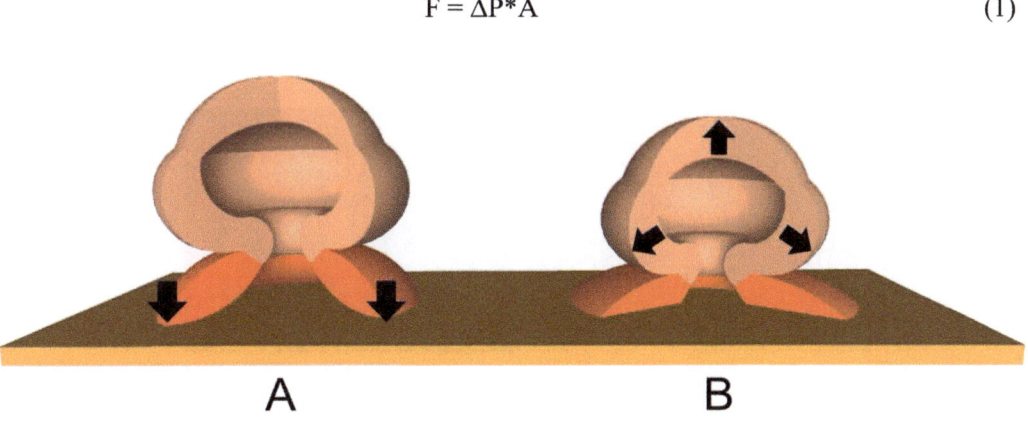

Figure 3 *Adhesion hypothesis: A) the infundibular part of the sucker creates a seal with the substrate at the rim (black arrows); B) the acetabular radial muscles contract (black arrows) to reduce the pressure inside the sucker, resulting in attachment to the susbstrate[18-19]; more specifically, the muscles exert a force on the water inside the acetabular cavity, which balances this force because of incompressibility, reducing the internal pressure of the cavity.*

Thus, the octopus sucker increases the attachment area by using roughness and grooves; and the pressure difference by exploiting the ability of water to sustain a tensile stress, maximising both the attachment area and the pressure difference.

Another interesting strategy that has been hypothesised as being used by the octopus is the storage of elastic energy in cross connective tissue fibres. The cross connective fibres are theorised to take the place of the contracting acetabular radial muscles in adhesion over extended periods of time. In this way, the octopus can attach without expending muscular effort[18-19].

4   METHOD AND RESULTS

3.1 Morphological Investigation

To better understand the octopus sucker structure, we performed a variety of morphological analyses (i.e., magnetic resonance imaging (MRI), ultrasonography, histology and Cryo-SEM)[24]. The information obtained from these different techniques was then compared and integrated (Figure 4).

We used histology to study the morphology and classify the different types of tissues. We used ultrasonography and magnetic resonance imaging (which are based on the external application of ultrasound waves and magnetic resonance fields, respectively) to

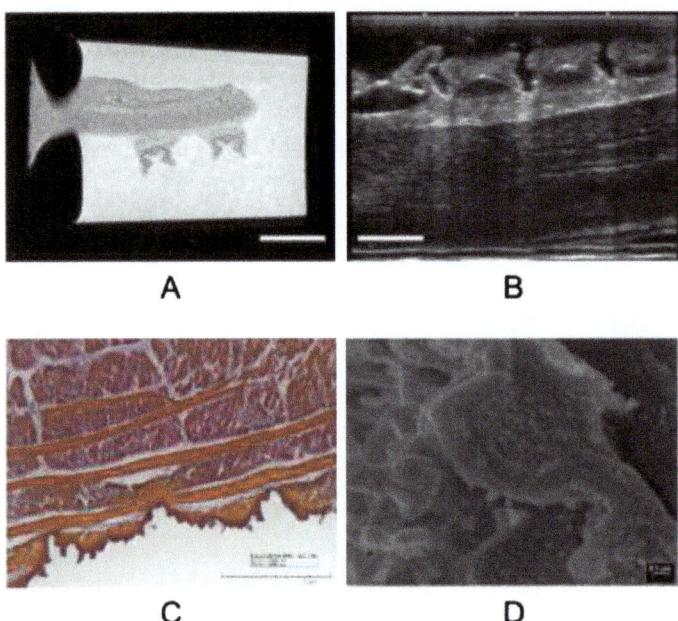

Figure 4   *Transverse section of an octopus sucker: A) magnetic resonance image of an octopus arm: the scale bar corresponds to 1 cm; B) ultrasonographic image of an octopus arm: the scale bar corresponds to 1 cm; C) histological image of infundibular circular muscles; note the transverse section showing the grooves and roughness: the scale bar corresponds to 1 mm; D) cryo-SEM image of a denticle, the elementary unit of infundibular roughness: the scale bar corresponds to 0.5 μm.*

Table 1 *Summary of the techniques used in morphological investigations of* Octopus vulgaris *suckers; the columns correspond to the following: details, i.e., machine/technique used; samples, i.e., the type of samples taken, in particular if the samples were collected ex vivo or in vivo and if they treated for analysis; and resolution, i.e., the spatial resolution of technique*

TECHNIQUE	DETAILS	SAMPLES	RESOLUTION
MRI	1.5 T MRI (Philips Achieva)	*ex-vivo* *Not treated*	mm
ULTRASONOGRAPHY	22 MHz Imager (Esaote MyLab Five Vet) with 20 mm linear array (SL3116)	*in-vivo* *Not treated*	1/10 mm
HISTOLOGY	Picro Ponceau	*ex-vivo* *Treated*	1/100 mm
CRYO-SEM	Cryo (SCU 020 Bal-Tech) dedicated to a SEM (Philips SEM 515)	*ex-vivo* *Treated*	several nm

avoid altering the structure of the biological tissues. These techniques are commonly exploited as human diagnostic methodologies. Recently, ultrasonography has been used to investigate the octopus brain and arm[25-27] because no damage is inflicted on the biological tissues. Moreover, ultrasonography facilitates non-invasive ultrasonographic recordings *in vivo*. We employed Cryo-SEM to examine structures in millimetric samples that could be not visualised by other techniques. The anatomical structure of the octopus suckers was investigated using MRI and Cryo-SEM for the first time in the present work. The main features of each technique are summarised in Table 1.

3.2 3D Reconstruction of the Octopus Sucker Structure

3D reconstructions were realised from the histological and magnetic resonance images using AMIRA software (Konrad-Zuse-Zentrum für Informationstechnik, Berlin, Germany) (Figure 5). Thus, CAD models of octopus suckers were obtained to use as a starting point

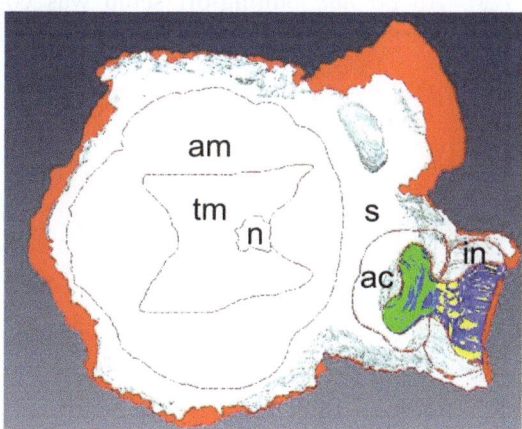

Figure 5 *3D reconstruction of an octopus arm from histological images: arm muscles, am; arm transverse muscles, tm; nerve cord, n; skin, s; acetabulum, ac; infundibulum, in.*

for designing a perfect artificial reproduction. The CAD model possessed the same features as the natural counterpart in terms of size and proportions.

3.3 Passive Prototypes: Reconstruction of the Infundibulum

We used careful morphological analysis[24] and 3D reconstructions to design a CAD model of the octopus sucker, from which we built the first soft passive prototypes. These prototypes were made of silicone and were tested using an *ad hoc* adhesion setup with external actuation[28].

   3.3.1 Fabrication of Passive Prototypes. The passive suction cups were fabricated by casting silicone rubber into a mould modelled on the octopus morphological data. The mould consisted of two specular portions that modelled the external surface of the prototype and a core that reproduced the acetabular internal chamber, the rim and the infundibular grooves (Figure 6). All the mould components were built in resin by rapid prototyping (3D System ProJet HD3000, USA).

   The prototypes were made of a soft elastomeric material that is strong and able to stretch to many times its original size without tearing and rebound to the original form without distortion. The particular material chosen was Dragon-Skin (from Smooth-On, USA). This material is very compliant and has been already used for building artificial soft components in robotic applications[29-30].

   The silicone rubber was made by mixing two liquid components for one minute, followed by degassing for two minutes in a vacuum oven. The resulting compound was poured into the two specular portions of the mould, the core was inserted and the parts were quickly closed. After eight hours of a curing phase at room temperature, the specular portions were separated and the prototype was released with the core still inserted. The core was then carefully removed from the prototype, exploiting the elastic properties of the rubber material by sliding the core through the orifice.

   The morphology of the developed prototypes was visualised with an optical digital microscope (HIROX KH-7700) (Figure 7).

   3.3.2 Adhesion Test. An *ad hoc* pulling-off setup was constructed to test the attachment capability of the prototypes and their subsequent compliance with different

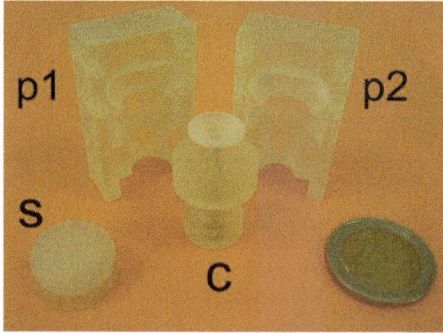

Figure 6   *Mould components: specular parts for fabricating the external form of the suction cup, p1 and p2; core for recreating the acetabular cavity and the infundibular morphology, c; and suction cup prototype, s. A 2 euro coin is shown at bottom right.*

Figure 7   *A) Optical microscope image showing the surface of the infundibular prototype surface: note the radial grooves, g; and the rim, r. B) Photograph of the developed prototype with a 1.5 cm diameter infundibulum and a weight of 2 g: a 50 cent euro coin is also shown.*

substrates, similarly to octopus suckers. Being passive at this stage, the prototypes were activated by an external suction system. The general aim of this analysis was to measure the pulling-off force needed to detach the suction cup from different substrates in water by using an external syringe to create suction.

The experimental apparatus was composed of anchoring, suctioning and pulling-off systems (Figure 8). The anchoring system consisted of a manual positioning stage (f), a water container (e) and a clamp unit (d). A water container was placed on the manual positioning stage and the clamp unit was fixed to the water container. The clamp unit held the suction cup in place, exposing the infundibular surface. This configuration allowed the suction cup to maintain its shape during the pulling-off tests, preventing distortion that could interfere with the measurements. The suctioning system consisted of a syringe (m)

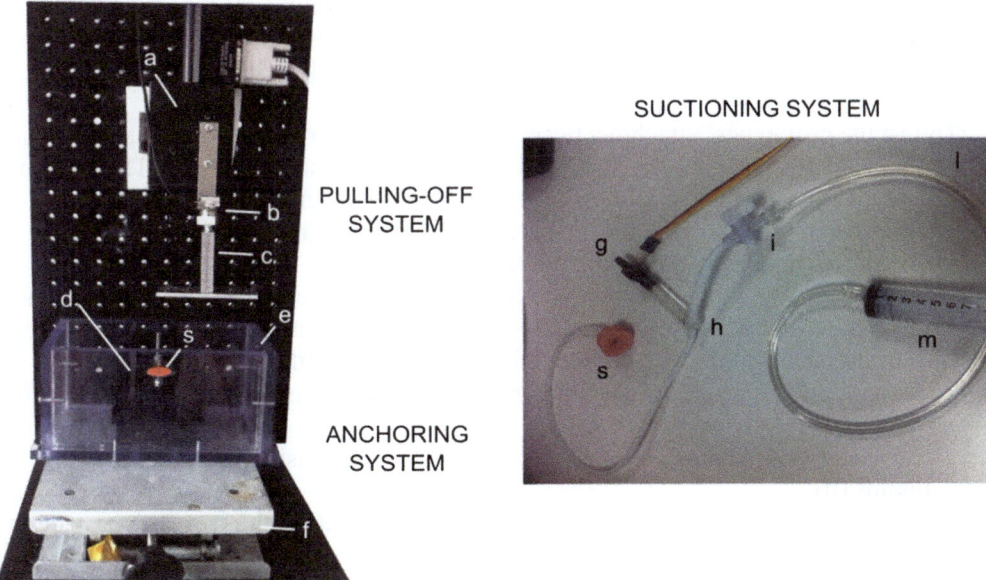

Figure 8   *Experimental apparatus with z-axis controlled translation stage, a; six-component load cell, b; T-shape link, c; clamp unit, d; water container, e; manual positioning stage, f; differential pressure sensor, g; T-joint, h; stopcock, i; tube, l; syringe, m; and suction cup prototype, s.*

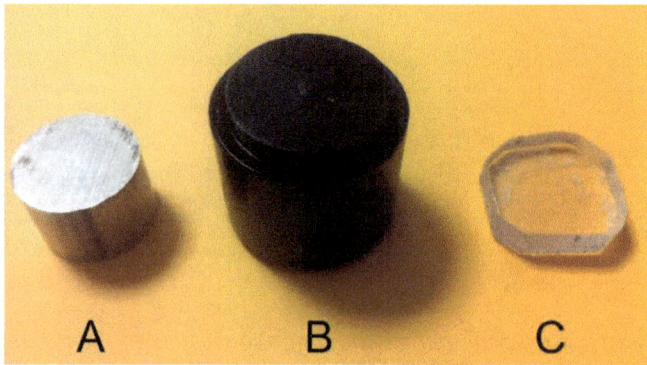

Figure 9   *Samples of substrates tested: A) aluminium sample with an average roughness of 9.660 μm and a peak roughness of 50.856 μm, μ1; B) Delrin® sample with an average roughness of 0.736 μm and a peak roughness of 5.269 μm, μ2; and C) Plexiglas® sample with an average roughness of 0.054 μm and a peak roughness of 1.531 μm, μ3.*

(1.54-cm in diameter), plastic flexible tubes (l), a stopcock (i) and a differential pressure sensor (g). The tubes were used to link the syringe to the suction cup, while the differential pressure sensor was placed between the syringe and the suction cup using a T-joint (h). The stopcock was placed between the syringe and the T-joint to maintain a constant pressure inside the suction cup after suction, without applying an external force.

The pulling-off system consisted of a micrometric translation stage with crossed roller bearings (M-126.CG1 PI, Karlsruhe, Germany) (a); a six-component load cell (ATI NANO 17 F/T, Apex, NC, USA) (b); and a T-shape link for placing the substrate samples (c). The data from the load cell, which corresponded to the normal pulling-off force, were acquired at a frequency of 1 kHz and an averaging level of 16. The signal from the differential pressure sensor was obtained using a National Instruments (NI) USB6009 and an NI Signal Express at a frequency of 20 Hz.

To test the prototypes' attachment capability, we used three substrates that were made of different materials and characterised by different roughnesses (Figure 9). The roughness was measured with a surface roughness profiler (Zeiss TSK Surfcom 130A).

The following protocol was used to perform the experimental test:
1. Fix the substratum to the T-shape link.
2. Bring the infundibular portion of the prototype into contact with the substratum using the manual positioning stage. All the elements must be immersed in water.
3. Start data acquisition for both the pulling-off force and the pressure difference.
4. Withdraw the syringe to exert a force on the water inside the suction cup (t0 in Figure 10).
5. Close the stopcock to maintain a constant pressure inside the suction cup (t1 in Figure 10).
6. Detach the suction cup from the substrate by activating the translation stage at a velocity of 0.3 mm/s in the z-direction (t2 in Figure 10).
7. Stop data acquisition when the sucker has detached from the substrate (t3 in Figure 10).

The experimental procedure was performed with care taken to exclude air bubbles from the suction system.

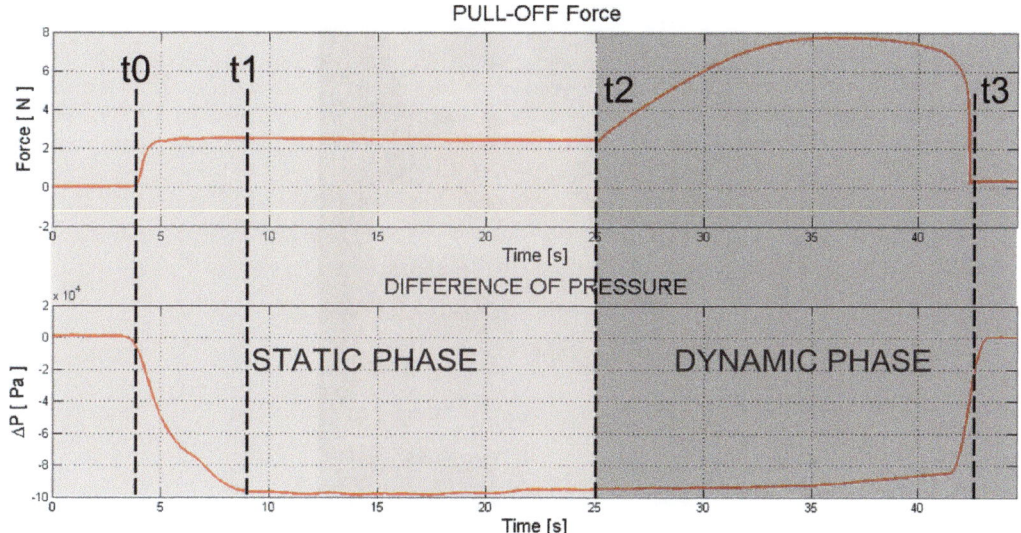

Figure 10 *Typical experimental profiles are shown above: the abscissa represents the acquisition time in s and the ordinates represent the normal pulling-off force in N and the pressure difference in Pa, respectively. For t < t0, the suction cup prototype is brought into contact with substratum; at t = t0, the syringe is withdrawn to create a suction; at t = t1, the stopcock is closed to maintain a constant pressure inside the suction cup; at t = t2, the translation stage is activated in the z-direction to detach the suction cup from the substrate; and at t = t3, detachment occurs. During the static phase, the suction cup and substrate are in contact but there is no motion (the translation stage is not yet activated); during the dynamic phase, the substrate is moved away from the suction cup in the z-direction (the translation stage is activated). We observe that when suction is applied during the static phase, the pressure difference increases (in absolute value) and the first pulling-off force is recorded; when the stopcock is closed instead, both of the variables remain constant at their most recent values. During the dynamic phase, when the translation stage is activated, the pulling-off force significantly increases. This force reaches its maximal value and then begins to decrease as the pressure difference also begins to decrease. This decrease occurs because the seal between the suction cup and the substrate is weakened by the detachment process. When detachment occurs, the pulling-off force and the pressure difference return to zero.*

3.3.3 Results. The developed prototypes showed good attachment capability to different types of substrates in terms of materials and roughness under wet conditions.

Figure 11 shows the adaptability of the suction cup prototype to a 200 g cylindrical aluminium object (which is 30 mm in diameter). In this case, suction was created using a syringe inserted into the upper part of the prototype.

A representative experimental result showed that a suction cup prototype with an outer diameter of 1.5 cm and a weight of 2 g had a load capacity of almost 8 N for an externally applied pressure difference of 10^5 Pa. These results were rather satisfying compared to

Figure 11 *Qualitative test on a suction cup for lifting a 200 g cylindrical aluminium object.*

other bio-inspired prototypes (which exhibited a load capacity of almost 2 N)[17]. Figure 12 shows the results obtained for the three types of substrates.

3.4 Active Prototypes: Long-term Attachment Strategy

The hypothesis that octopus suckers can perform adhesion for extended periods of time without any muscular effort has inspired the design of the first prototype of an artificial suction cup that can perform efficient adhesion over a long time[31].

The active suction cup prototype consists of two main parts, analogous to octopus suckers: an acetabular chamber (ac) and an infundibular portion (in) (Figure 13). Embedded in the acetabular chamber is a sliding piston (p) made of a hard material (e.g., resin) that acts as the suction unit. The infundibular portion is soft (e.g., silicone) and is compliant to the external surface. Consequently, the infundibular portion can be realised following the same strategy used for the passive suction cup prototype. In this way, the artificial infundibulum can exploit the strategic morphology of biological suckers (namely, the radial grooves and the rim), which we have already developed as passive prototypes.

The acetabulum and infundibulum are connected by an orifice that includes a valve (v) made of resin. The valve consists of a drilled disk (d) and a support (s); the disk radial periphery contains holes and the disk can slide vertically into the support (Figure 14). The support is designed so that the holes are open (up-phase) when the drilled disk is in contact with the upper face of the support (Figure 14 A-B); when the drilled disk is in contact with the lower face of the support, the holes are closed (down-phase) (Figure 14 C-D).

The procedure for obtaining long-term adhesion and minimising external forces using the design described above consists of the following main steps:

- Bring the infundibular portion into contact with the substratum and then compress the infundibular portion against the substratum;
- Withdraw the piston of the acetabular chamber to exert a force on the water inside the device, thereby inducing a pressure drop. Following the piston withdrawal, the valve is passively brought into the up-phase position and the holes in the drilled disk allow water to flow from the infundibular section to the acetabular section;

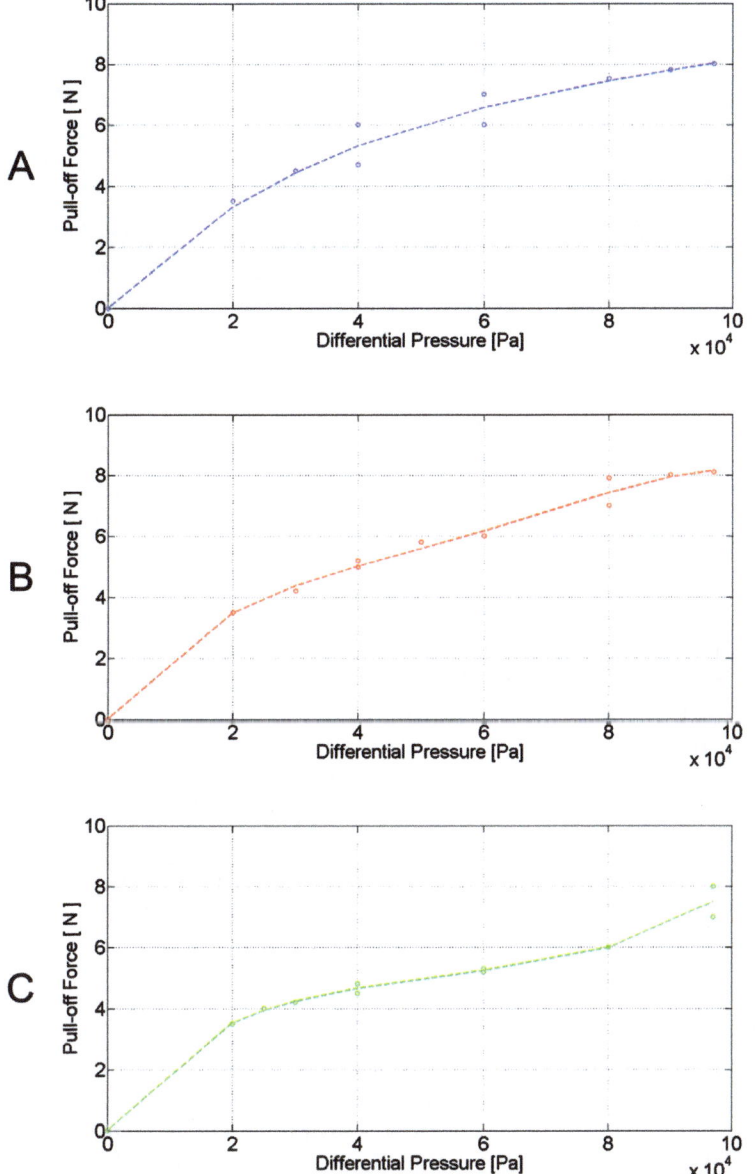

Figure 12 *Graphs showing the pulling-off force in response to decreasing the pressure inside the suction cup: A) aluminium substratum (µ1); B) Delrin® substratum (µ2); and C) Plexiglas® substratum (µ3) (see Section 3.3.2). The graphs show that the suction cup prototypes were able to resist almost 8 N of pulling-off force for an applied pressure difference of 10^5 in all three cases, although different trends were observed in each case.*

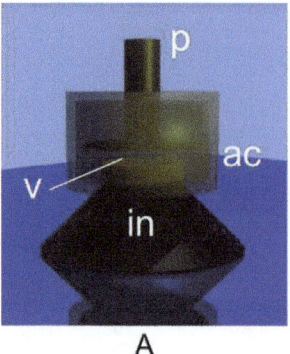  

A   B   C

Figure 13 *A) Design of an active suction cup prototype: piston,* **p***; acetabular chamber,* **ac***; valve,* **v***; and infundibular portion,* **in***; B) shows that the infundibular portion, being made of a soft material, can be compressed onto the substratum, creating a seal against the environment; C) a sliding piston, embedded in the acetabular chamber, can exert a force on the water inside the suction device.*

-Bring the valve into the down-phase position and then remove the suction generator (by releasing the piston). Having closed the valve to remove the suction generator, the pressure in the acetabulum increases, while remaining low in the infundibular section (because in the down-phase position, the holes in the drilled disk, which facilitate flow between the two compartments, are completely occluded).

In such a configuration, the valve is passively maintained in the down-phase position by the pressure difference created between the acetabulum and the infundibulum, resulting in adhesion over an extended period of time without any energy consumption.

The prototype proposed above exploits the particular morphology of the infundibular section and a long-term attachment strategy without energy consumption, which were both inspired by the octopus sucker.

4 CONCLUSIONS

The first adhesion tests performed with passive suction cups have demonstrated good device attachment capability onto different substrates. These results are an important benchmark for the adhesion strength of artificial devices inspired by the morphology of octopus suckers. The current limiting factor in these experiments is the inability to induce a negative pressure within the acetabular chamber, as occurs in octopus suckers. This problem is caused by the cavitation threshold. Cavitation occurs due to the formation of vapour bubbles within fluids due to extreme internal pressure reduction. This bubble formation compromises device attachment and prevents higher tensile stresses from being induced. This phenomenon depends on different factors, such as the liquid type, the type of substrates and materials involved in the adhesion process, the presence of impurities and the temperature. Possible solutions are to use a more compact setup (without a T-joint, limiting the tube length and so on in order to reduce some nucleation sites for cavitation bubbles), to control the temperature, to clean the substrates more carefully and so on. A higher water tensile stress could thus be induced to increase the adhesion strength of our suction cup prototypes to a value comparable to that of the octopus sucker.

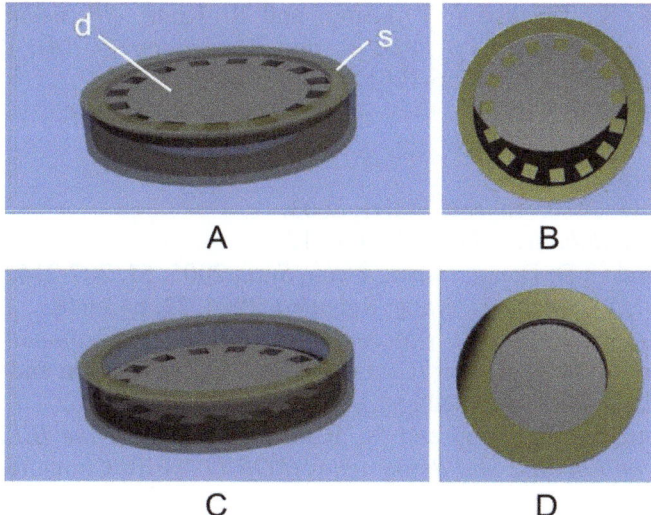

Figure 14 *Valve design showing drilled disk, d and support, s: A) drilled disk in contact with the upper face of the support where the holes of the drilled disk are open (up-phase); B) top view of A; C) drilled disk in contact with the lower face of the support where the holes of the drilled disk are closed (down-phase); and D) bottom view of C.*

Given that the lowest negative pressure reported in the octopus sucker is approximately -0.168 MPa[32-33], an artificial sucker mimicking this natural behaviour, with an adhesion area of 1 cm, for example, could attain a 21 N adhesion force.

Moreover, another important development of our passive prototype is the use of a rough infundibular surface to increase the adhesion capabilities of the developed devices.
Next, the strategies of the passive and active prototypes need to be integrated to mimic the full capability of the biological counterpart. In addition to improving the infundibular features, we particularly want to embed into the prototype a suction unit (represented by a piston thus far) and a valve system that guarantees efficient adhesion over an extended period of time.

A design realising all of these objectives should usher in a new generation of efficient artificial suction cups.

References

1. *US Pat.*, US4582353, 1986.
2. Y. Yoshida and M. Shugen, *Proceeding of the IEEE International Conference on Robotics and Biomimetics (ROBIO)*, IEEE Computer Society, Tianjin, 2010, pp. 1513-1518.
3. B. Bahr, Y. Li and M. Najafi, *Comput. Electr. Eng.*, 1996, **22**, 193-209.
4. J. Liu, K. Tanaka, L. M. Bao and I. Yamaura, *Vacuum*, 2006, **80**, 593-598.
5. T. Miyake, H. Ishihara and T. Tomino, *Proceeding of the IEEE International Conference on Robotics and Biomimetics (ROBIO)*, IEEE Computer Society, Guilin, 2009, pp. 1824-1829.

6. J. Berengueres, K. Tadakuma, T. Kamoi and R. Kratz, *Proceeding of the IEEE International Conference on Robotics and Automation (ICRA)*, IEEE Computer Society, Roma, 2007, pp. 1256-1261.
7. *EP Pat.*, EP1557083B1, 2010.
8. *US Pat.*, US7878467, 2011.
9. *US Pat.*, US5345935, 1994.
10. *US/EP Pat.*, US6723105/EP0923348B1, 2003.
11. Z. Xu and P. Ma, *Robotica*, 2002, **20**, 209-212.
12. D. M. Aslam and G. D. Dangi, *Robot. Auton. Syst.*, 2005, **51**, 207-214.
13. Y. Zhao, Z. Fu, Q. Cao and Y. Wang, *Robotica*, 2004, **22**, 643-648.
14. W. Brockmann and F. Mösch, *Proceeding of the 7th International Conference on Climbing and Walking Robots (CLAWAR)*, eds. M. A. Armada and P. González de Santos, Springer Berlin Heidelberg, Madrid, 2004, pp. 935-942.
15. Z. Tao, L. Rong, X. D. Wang and W. Kun, *Proceeding of the IEEE International Conference on Robotics and Biomimetics (ROBIO)*, IEEE Computer Society, 2006, pp. 491-495.
16. A. Packard, in *The Mollusca-Form and Function*, eds. E. R. Trueman and M. R. Clarke, Academic Press, San Diego, 1988, vol. 11, pp. 37-67.
17. B. s. Hu, L. w. Wang, Z. Fu and Y. z. Zhao, *Int. J. Adv. Robotic Sy.*, 2009, **6**, 151-160.
18. W. M. Kier and A. M. Smith, *Biol. Bull.*, 1990, **178**, 126-136.
19. W. M. Kier and A. M. Smith, *Integr. Comp. Biol.*, 2002, **42**, 1146-1153.
20. W. M. Kier and K. K. Smith, *Zool. J. Linn. Soc.-Lond.*, 1985, **83**, 307-324.
21. P. Girod, *Arch. Zool. Exp. Gen.*, 1884, **2**, 379-401.
22. A. Naef, in *Fauna and Flora of the Bay of Naples*, Israel program for scientific translations, Jerusalem, 1921-1923, vol. 35, pp. 1-917.
23. M. Nixon and P. N. Dilly, *Symp. Zool. Soc. Lond.*, 1977, **38**, 447-511.
24. F. Tramacere, L. Beccai, E. Sinibaldi, C. Laschi and B. Mazzolai, *Procedia Computer Science*, 2011, **7**, 192-193.
25. A. M. Grimaldi, C. Agnisola and G. Fiorito, *Brain Res.*, 2007, **1183**, 66-73.
26. L. Margheri, B. Mazzolai, M. Cianchetti, P. Dario and C. Laschi, *Proceeding of the Annual International Conference of the IEEE Engineering in Medicine and Biology Society (EMBC)*, eds. B. He, Y. Kim, X. Pan and G. A. Worrell, IEEE Computer Society, Minneapolis, 2009, pp. 7196-7199.
27. L. Margheri, G. Ponte, B. Mazzolai, C. Laschi and G. Fiorito, *J. Exp. Biol.*, 2011, **214**, 3727-3731.
28. F. Tramacere, L. Beccai, F. Mattioli, E. Sinibaldi and B. Mazzolai, *Proceeding of the IEEE International Conference on Robotics and Automation*, IEEE Computer Society, St. Paul, 2012, pp. 3846-3851.
29. M. Cianchetti, A. Arienti, M. Follador, B. Mazzolai, P. Dario and C. Laschi, *Mater. Sci. Eng.: C*, 2011, **31**, 1230-1239.
30. C. Laschi, B. Mazzolai, M. Cianchetti, L. Margheri, M. Follador and P. Dario, *Adv. Robotics*, 2011, **26**, 709-727.
31. F. Tramacere, L. Beccai and B. Mazzolai, *Lect. Notes Comput. Sc.*, 2012, **7375**, 400-401.
32. A. M. Smith, *J. Exp. Biol.*, 1991, **157**, 257-271.
33. A. Smith, *J. Exp. Biol.*, 1996, **199**, 949-958.

MUSSEL-INSPIRED ADHESIVE INTERFACES FOR BIOMEDICAL APPLICATIONS

H. Ceylan, A.B. Tekinay* and M.O. Guler*

Institute of Materials Science and Nanotechnology, National Nanotechnology Research Center (UNAM), Bilkent University, Ankara 06800, Turkey.
*atekinay@unam.bilkent.edu.tr, moguler@unam.bilkent.edu.tr

1 INTRODUCTION

Underwater adhesion is a daunting task for organisms living in the intertidal zones. Irregularities in salinity, abrasive wearing of the ocean waves, microbial invasions, and sharp thermal fluctuations create an environment of harsh extremes. On the other hand, continuous supply of nutrients moved from the ocean to the living zone of the organisms through the tides sustains an ecosystem inhabited by unique organisms, some of which exhibit exceptional adaptive characteristics for adhesion under highly unstable conditions. Adhesion strategies of different organisms living in these zones, such as sandcastle worm, adult barnacles and their larvae (the cyprid), and mussels have been under close examination by researchers in recent years with the purpose of translating their adhesive technology into synthetic platforms for industrial and medical applications.[1-10] A natural underwater adhesive is usually synthesized and secreted onto the substrate with low initial viscosity followed by curing of the glue over the course of minutes to hours into its final hardened structure. An impressive feature of these adhesives is that they require little or no surface preparation prior to secretion.[11] This capability is desirable in man-made adhesives which, so far, are incapable of functioning on highly solvated surfaces. Highly polar water molecules, dissolved ions and organic contaminants interact both with the surface and the adhesive molecules, interfering with the adhesion process.

The nature of biofunctional interface, separating an inert biomedical device from the native tissue while integrating the material into the body, is of utmost importance for the long-term efficiency of tissue regeneration. In order to achieve this, strong and biologically safe underwater synthetic adhesives, which can modulate cellular activities through biologically active signals, are required.[12-14] The mainstream research in this field has largely concentrated on creating artificial cellular microenvironments by mimicking the architecture and biology of the native extracellular matrix. Toward this purpose, various polymers and self-assembled nanofibers of peptidic structures have been engineered to present desired biofunctional ligands to interfere with cellular signalling. On the other hand, immobilization of such materials onto a substrate has remained an unresolved issue. Physical and chemical conditions in the applications where these adhesives are normally used both resemble the conditions in the intertidal zone. For example, the mechanical abrasiveness in load-bearing tissues, such as bone and cartilage, the high shear force experienced in blood vessels and the high ionic strength and polyionic environment of bodily fluids create a challenging environment for adhesives to operate efficiently. Therefore, natural underwater adhesives provide a plethora of inspiration towards developing biologically safe and reliable synthetic adhesives for medical applications.

The focus of this chapter is to provide a perspective of underwater adhesion in nature and biomimetic attempts to reconstitute natural strategies in synthetic platforms. The adhesion system of the mussel is given special priority because of the extensive biomimetic research on this subject, and an overview of mussel adhesion from materials science perspective is described in section 2. Many excellent reviews previously discussed mussel adhesion mechanism in detail.[15-17] In section 3, recent and seminal biomimetic attempts for functionalization of surfaces from nano to macro levels are reviewed. In section 4, the discussion is centred upon a modular concept towards reconstitution of a mussel-mimetic, polyphenolic, supramolecular adhesive at physiological pH, with the purpose of creating cell-guiding microenvironments to enhance biocompatibility of metal implant surfaces.[18-19]

2 AN OVERVIEW OF MUSSEL ADHESION AS SOURCE OF INSPIRATION

Underwater adhesion of mussels, particularly *Mytulis edulis and Mytulis californianus*, has drawn escalating interest in biomimetics research.[16-17] In their natural habitat these sessile (non-motile) organisms cling to underwater solid surfaces (e.g., rocks, wood, etc.) in the intertidal zones. The adhesion capacity of mussels encompasses virtually all types of surfaces, including metals, alloys, metal oxides, organic surfaces, and plastics, even polytetrafluoroethylene (TEFLON®).[26-27] Due to their extensive adhesion capacity, mussel fouling of ship hulls and coastal infrastructure has been a growing economic concern. There is no man-made glue that can bind to such a broad variety of surfaces. The adhesion strength of mussels is one of the strongest known in natural underwater adhesives.[17] In order to achieve this, mussels produce a special polyphenolic glue containing hierarchically organized proteins, known as mussel adhesive proteins (mfps), with varying content of 3,4-dihydroxy-L-phenylalanine (DOPA) residues (Figure 1 and Table 1). Spatial and temporal evolution of this phenolic residue within the wet glue precursor is believed to play an indispensible role in mussel adhesion and cohesion. The mussel adhesive unit is called a byssal, or byssi in plural, thread, or byssus in plural, and contains three main functional and biochemical components: a stem embedded in the soft tissues of the mussel, a hard and flexible thread-like extension (byssal thread), and an adhesive plaque (Figure 1).[16,28] The whole structure is composed of different proteins called mfps. Up to now, six major proteins have been isolated and identified with different highly organized functions in the adhesion process (mfp-1 to mfp-6).

The catechol side chain of DOPA can be involved in a variety of physical and chemical reaction mechanisms. For example, catechols can form exceptionally stable complexes with metals and metal oxides, thereby mediating adhesion to these surfaces. Lee et al. measured the dissociation force of a single tethered DOPA molecule from TiO_2 using single-molecule atomic force microscopy (AFM).[25] The dissociation force of the single DOPA-TiO_2 bond (~800 pN) is about half of a covalent bond (~2000 pN) and much higher than the dissociation force of hydrogen bonds that hold the DNA double helix intact (10-20 pN).[25,32-33] In spite of the high bond strength, DOPA binding to TiO_2 surface is completely reversible with thousands of break/reformation series.[25] Catechol groups in mfps can also undergo covalent reactions that contribute significantly to the cohesive and water-resistant characters of the mussel glue.[34-35] Under the basic conditions of seawater (pH~8.5),

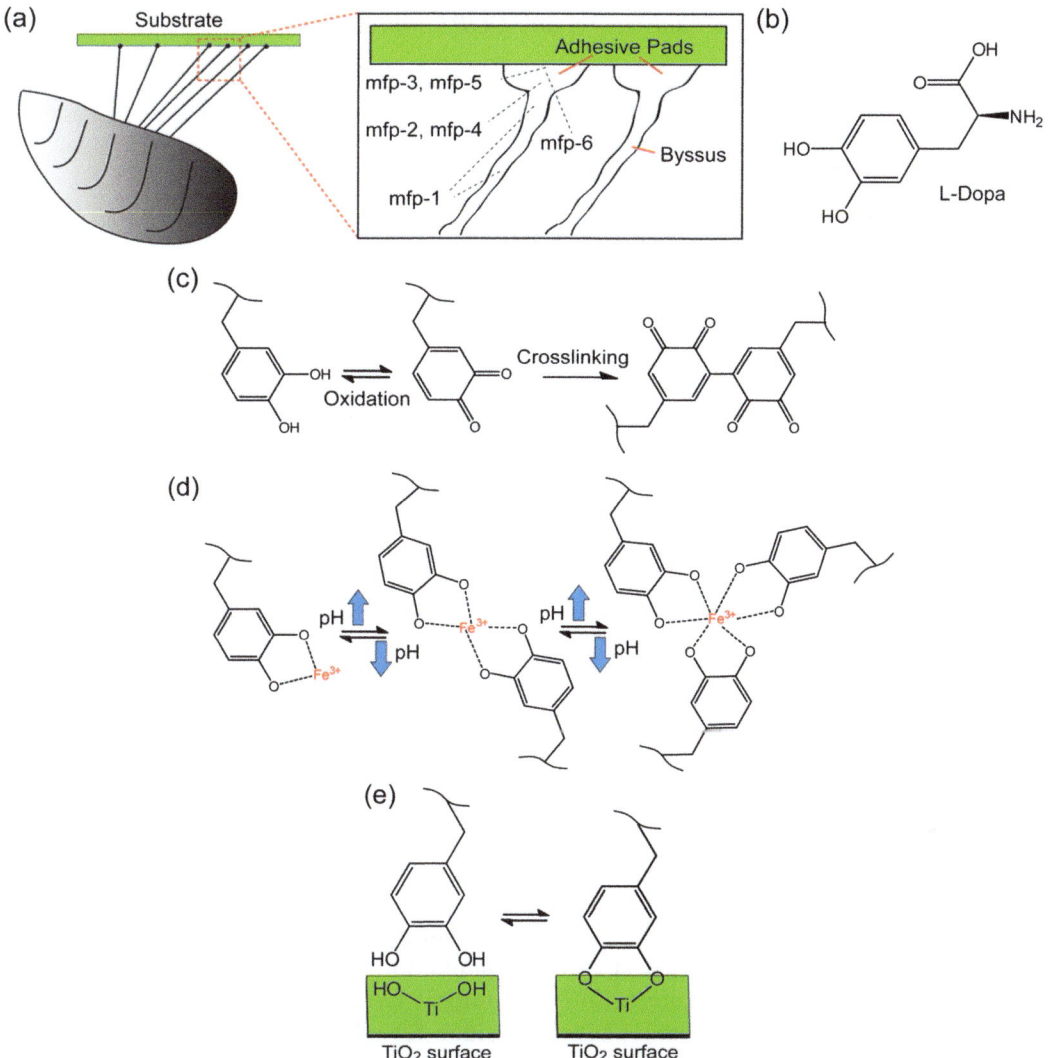

Figure 1 *(a) A sketch of mussel adhered to a substrate through its byssal threads and adhesive plaques. Mussel glue is composed of hierarchically organized mfps with varying DOPA content. (b) Chemical sketch of L-DOPA. (c) Oxidation-mediated covalent crosslinking of DOPA residues in mfps.[20-21] (d) pH-dependent reversible complexation of mfps with ferric iron ions.[22-23] (e) Reversible binding of DOPA onto TiO₂ surface.[24-25]*

catechol is oxidized to highly reactive quinone and semiquinone species that further react with each other to covalently crosslink (cure) the mfps.[27,35] Further, DOPA-quinone was reported to covalently react with primary amine and sulfhydryl groups.[25,36]

A byssal thread is composed of an extensible inner core and a hard outer shell (the cuticle).[15] Typically, hardness of the cuticle is roughly five times higher than the hardness of inner core proteins.[28] The inner core is formed by extracellular matrix proteins, with a central collagen core flanked by silk and elastin-like domains.[37] Because of the spatial

Table 1   *Molecular weight and DOPA content of mfps.*[7, 29-31]

Proteins	Molecular weight (kDa)	DOPA content (mol %)
mfp-1	108	10-15
mfp-2	46	3
mfp-3	6	21
mfp-4	80	5
mfp-5	8.9	27-30
mfp-6	11.6	4

distribution of the silk- and elastin-collagen complexes in the thread, the distal portion (closer to the substrate) is typically an order of magnitude more rigid than the proximal (closer to the mussel) portion. However, the distal and proximal portions are extensible up to 109 and 200%, respectively, without breaking apart.[38-40] The cuticle consists of densely packed granules, 0.8 µm in diameter, constituting ~50% of the cuticle volume.[15] mfp-1 is a 108 kDa structural protein containing a unique repetitive decapeptide, and it coats the entire adhesion plaque and the distal portion of the byssus.[41] Recent evidence suggests that the granules in the cuticle are formed by mfp-1 proteins densely crosslinked by ferric iron ions. mfp-1 can form reversible bis-Fe(DOPA)$_2$ and tris-Fe(DOPA)$_3$ complexes with iron ions in alkaline sea water.[28] These complexes have very high stability constants (log K_s ~37-40), implying that iron ions are a critical element of mussel adhesion that endows the cuticle with both hardness and self-healing ability after fracture.[42] Holten-Anderson et al. showed that removal of iron ions largely inhibits self-healing of DOPA-conjugated polymer and supramolecular networks.[22] Despite its high stiffness, the cuticle is highly extensible, with ultimate tensile strain around 70%.[43] Deformations up to this point are prevented from propagating by densely-crosslinked granules, whereas micro tears that are formed during deformation are mended by a rapid healing process.[28, 43]

Adhesive plaques establish adhesion of the animal to the surface.[7,44] Proteins that are in direct contact with the substrate (mfp-3 and mfp-5) have the highest DOPA content, highlighting the significance of this phenolic residue for adhesion (Table 1). mfp-3 and mfp-5 both contain a high number of cationic arginine and lysine residues, respectively.[16,29] The ε-amine group of lysine is reactive with oxidized catechol (quinone), which may entail crosslinking in the mussel glue.[36] However, *ex vivo* studies of mussel proteins and DOPA-containing peptides have not yet confirmed such reactivity.[20-21] The current view is that the excess positive charge forms Coulombic interactions with surfaces that mussels bind to in their native environment, such as rocks that are rich in negatively charged silicates and aluminates.[45-47] mfp-5 is also enriched in serine and phosphoserine residues, whose roles remain unknown.[5] Although, mfp-3 and mfp-5 are more enriched with DOPA, the bulk of the adhesive plaque largely consists of mfp-1, mfp-2, and mfp-4. mfp-2 is known to be resistant to proteolysis and is thought to act as the stabilizer of byssus cement.[44] This protein contains a large number of cysteine residues. mfp-4, on the other hand, is thought to serve as a bridge in the thread-plaque junction by linking the core collagen fibers of the distal byssus to plaque proteins, and has high levels of histidine, arginine, and lysine.[48-49] mfp-6, which was discovered in *Mytilus californianus*, has surprisingly low DOPA (4%) and high cysteine content (11%).[31] Zhao and Waite reported the presence of 5-S-Cysteinyl DOPA, a cysteine-DOPA adduct. Therefore, this protein is believed to provide a cohesive link between the surface-coupling proteins (mfp-3 and mfp-

5) and the bulk plaque proteins (mfp-2 and mfp-4).[31] A very recent report, however, attributed a more fundamental role to this protein by suggesting that mfp-6 prevents auto-oxidation of DOPA to quinone in mfp-3 before adhering onto substrate.[50] Quinone formation dramatically reduces the adhesion capacity onto TiO_2 and mica surfaces, by about 80%.[25,51] Thiol groups on side chain of cysteine residues in mfp-6 successfully maintain DOPA by coupling oxidation of thiols to reduction of quinones.

3 MUSSEL-INSPIRED SURFACE FUNCTIONALIZATION STRATEGIES

Despite the high underwater adhesive performance of the mussel glue, the inimitable complexity and the hierarchical organization of its constituent proteins has restricted the practical use of this material. Even isolation of the individual mfps is a demanding task due to their labour-intensive and inefficient production yield. For instance, approximately 10,000 mussels are required to extract only 1 g of mfp-1; and the purity of this extract is not reliable due to high batch-to-batch variation.[52] In order to produce this protein at large scale, *E. coli* and *S. cerevisiae* have been used,[53-55] however, attempts to produce functional mfps failed mainly due to codon bias and small expression quantity.[54-56] Although there has been partial success in the expression of mfp-1 repetitive sequences in *S. cerevisiae*[54] and *E. coli*[55-56] using synthetic gene constructs, their adhesion profiles were found to be poor. In another study, Choi et al. took a recombinant approach by fusing domains of mfp-1 and mfp-5 with functional groups of extracellular matrix proteins, such as RGD and YIGSR in order to create cell-friendly coatings.[57]

However, mfp production through recombinant production is still an unresolved challenge due to lack of the post-transcriptional modifications, including formation of DOPA by hydroxylation of Tyr residues.

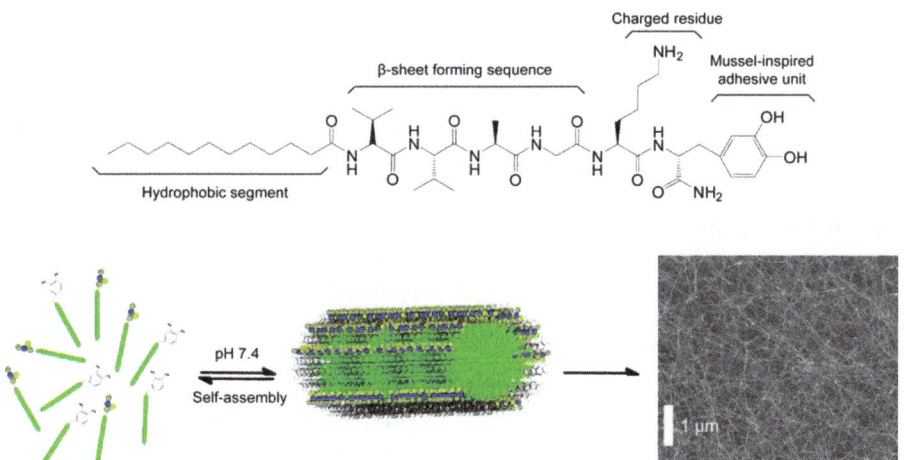

Figure 2 *Upper content: Chemical sketch of DOPA-modified peptide amphiphile (DOPA-PA) with modular domains required for self-assembly into nanofibers. Lower content: DOPA-PA can be mixed with any biofunctional peptide amphiphile to self-assemble into a nanofibrous adhesive network with desired biological activity.*

Limitations of obtaining high-purity and functional mfps led to alternative biomimetic approaches. Conjugation of DOPA, or the catechol group, to synthetic platforms has been the most widely recognized strategy. DOPA is chosen because not only can it adhere to a wide variety of substrates underwater, but also it has a very simple chemical structure, which can be easily grafted onto synthetic systems.[18-19,35] Use of a synthetic backbone with well-defined chemistry, onto which DOPA can be attached, offers a more reliable platform with minimum batch-to-batch variation compared to natural adhesive proteins. Moreover, biomimetic reconstitution of mussel adhesives on synthetic platforms with additional functionalities provides a wider range of applications and development strategies of novel hybrid materials with superior performance. Using this approach, catechol-conjugated poly(ethylene glycol) (PEG) was synthesized to obtain anti-fouling surfaces.[58-60] Likewise, catechol-functionalized chitosan/pluronic thermo-responsive and injectable hydrogels were utilized as tissue adhesives.[61] Using poly(dopamine methacrylamide-comethoxyethyl acrylate), a reversible dry/wet adhesive platform was developed by Lee et al. through combining mussel-mimetic DOPA adhesion with gecko-mimetic polydimethyl siloxane pillars.[62] DOPA-modified polymers were used to functionalize not only bulk surfaces, but also surfaces of nanoparticles. For instance, binding of methoxy poly(ethylene glycol), which was grafted to hyper-branched polyethylenimine and polyDOPA, onto hydrophobic nanoparticles provided stabilization in harsh biological environments.[63] Oxidation-mediated grafting of catechols conferred redox activity to chitosan films.[64] These films were characterized to be poor in direct electron transfer, whereas electrons can readily flow through soluble mediators. As a result of this interaction, catechol-modified chitosan films exhibited amplification, partial rectification, and switching capabilities, thereby holding promise for sensor development. Previously, we conjugated DOPA to self-assembling peptide amphiphile molecules and developed a mussel-inspired, supramolecular adhesive for metal implant biofunctionalization.[18-19] By co-assembling DOPA-bearing peptide amphiphile molecules with certain biofunctional ligand-bearing molecules, these bioactive coatings induced selectively enhanced endothelial cell and osteoblast activities on stainless steel and TiAl6V4 surfaces, respectively. The detail of this strategy is explained in the following section. In another study, Lee et al., took an elegant approach to mimicking adhesion between DOPA- and amine-rich mfp-3 and mfp-5 using a simple molecule, dopamine.[27]

At basic pH, dopamine undergoes an oxidation-triggered auto-polymerization reaction. Hong et al. revealed that both non-covalent assembly and covalent polymerization contribute to polydopamine formation.[65] Virtually any type of surface, regardless of its chemistry, can be coated with polydopamine by simply dipping it into dopamine solution at pH ~8.5. Moreover, polydopamine coating thickness is proportional to the time of immersion and the chemical properties of this coating allow secondary modifications through coupling to nucleophilies.[27,36] Ryu et al. demonstrated that the polydopamine coating provides a general route for bone-like hydroxyapatite crystallization on a surface.[66] Wei et al. functionalized superparamagnetic iron oxide nanoparticles (SPIONs) with a dopamine sulfonate ligand to provided nanoparticles stability in water against pH and salinity changes in addition to enabling further functionalization with streptavidin or a maleimide dye.[67] Very recently, Kang et al. proposed a one-step surface functionalization strategy by mixing dopamine with a diverse range of organic and inorganic species and dipping the substrate into this mixture.[68]

4 BIOFUNCTIONALIZATION OF METAL SURFACES THROUGH MUSSEL-INSPIRED POLYPHENOLIC SUPRAMOLECULAR NANOFIBERS

Reconstitution of artificial cellular microenvironments that direct cellular activities in a desired way has been an active area of research. When an implanted material (whether or not it is biologically inert) comes into contact with the body, a cascade of undesirable reactions takes place, often leading to failure of the tissue regeneration process and loss of the implant. In order to circumvent the biocompatibility issue, a promising strategy is to modify surfaces with biofunctional agents, so that native tissue will recognize the material as self and will grow on it. Native extracellular matrix regulates cellular behaviors in the cellular microenvironment by presenting cells spatially- and temporally-regulated biofunctional signals. Likewise, the artificial cellular microenvironment is engineered to provide cells with certain biofunctional ligands that guide cellular behaviors (e.g. adhesion, morphogenesis, viability, proliferation, migration and differentiation) for acceptance of the material into body and proper functioning of the regenerating tissue. On the other hand, stability of biofunctional coatings on the material surface is limited under the abrasive conditions of the body environment.

During the last decade, peptide amphiphiles (PA) have been one of the most widely used nanofiber platforms to reconstruct cellular microenvironments in both 2D and 3D for *in vitro* and *in vivo* applications.[69-72] Typically, the diameter of a PA nanofiber is ~10 nm and the mesh size of the entangled network is in the range of 5-200 nm.[73] In addition to providing mechanical support as a tissue scaffold, PA nanofibers present cells with instructive cues through biofunctional signals conjugated to the monomer building blocks, thereby guiding cells at the molecular level by directing cellular adhesion, spreading, proliferation and differentiation.[18-19, 72]

Basic PA structure contains several functional modules carrying information necessary to self-assemble into nanofibers in addition to chemical and biological agents, which can be densely presented when they assemble into nanofibers (Figure 2).[18-19,72] Typically, a hydrophobic segment attached to the N-terminus of a peptide sequence forces packing of PA monomers into micellar assemblies. In water, the hydrophobic segment is buried inside the nanofiber exposing hydrophilic part to the outer environment. Hydrophobic amino acids such as Val or Ala are usually conjugated next to the hydrophobic segment of the PA structure. The amide backbone of this part facilitates β-sheet secondary structure through hydrogen bonds in the direction of nanofiber elongation.

Table 2 *Sequences of DOPA-modified and biofunctional peptide amphiphiles described with targeted applications. The table was reconstituted from references 18 and 19.*

PA Sequence	Nomenclature	Net Charge*	Biofunctional Adhesives**	Targeted Application
Lauryl-VVAGK**DOPA**	DOPA-PA	+1	REDV-PA/DOPA-PA	Selective endothelialization of luminal stent struts
Lauryl-VVAGE**REDV**	REDV-PA	-3		
Lauryl-VVAGE**DOPA**	DOPA-PA	-1	KRSR-PA/DOPA-PA	Selective osteoblast growth on orthopedic implants.
Lauryl-VVAG**KRSR**	KRSR-PA	+3		

* Theoretical net charge at pH 7.4.
** Nomenclature is based on the participating PAs in the self-assembly.

Charged amino acids are preferably incorporated next to the hydrophobic ones in order to increase solubility and exploit their charged units as a switch for self-assembly. A biofunctional polypeptide sequence, such as Arg-Gly-Asp-Ser, or a wide range of chemical groups of interest, such as biotin can be presented on the nanofibers as guiding cues for cells.[74-75]

To form mussel-mimetic peptide nanofibers, we conjugated DOPA to the C-terminus of a self-assembling PA (DOPA-PA) as shown in Figure 2.[19] DOPA-PA can co-assemble with biofunctional PAs into entangled networks of nanofibers where both catechol and biofunctional ligands are densely presented together to the outer environment. Exploiting catechol chemistry, these nanofibers can adhere onto any type of surface, where the nanofibers with biofunctional signals will create an inductive cellular microenvironment. In order to build biofunctional interfaces on biomedical implant surfaces, we designed different DOPA-PAs and biofunctional PAs targeted for specific applications, which are shown in Table 2.

A major limitation of cardiovascular stents is impaired endothelialization on the luminal surface of the stent struts. Lining the innermost layer of coronary arteries, endothelial cells are responsible for the proper functioning of blood vessels, regulating proliferation of smooth muscle cells (SMCs) and preventing platelet activation inside the vessels.

On the other hand, a delay or impairment in the recovery of endothelium after damage leads to re-closure of the vessels through uncontrolled proliferation of SMCs into the arteries. In order to promote adhesion and growth of endothelial cells on stainless steel, which is the most widely used material for manufacturing stents, we developed a hybrid stent coating model based on self-assembled nanofibers of REDV-PA and DOPA-PA.[19] The REDV ligand is recognized by $\alpha 4\beta 1$ integrins and were reported to selectively promote endothelial cell adhesion and spreading over smooth muscle cells and fibroblasts.[76] REDV-PA/DOPA-PA nanofibers could successfully adhere to stainless steel surfaces through catechol groups and this was confirmed by X-ray photoelectron

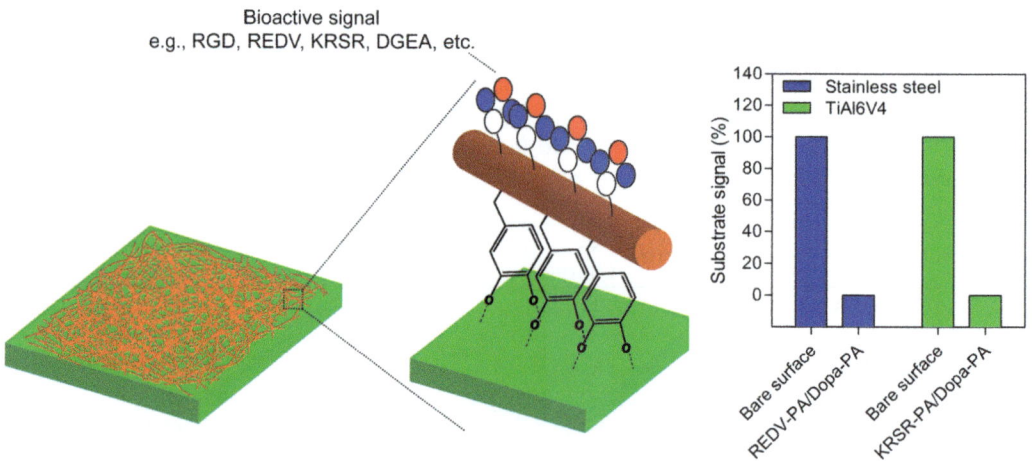

Figure 3   *A sketch showing nanofiber adhesion onto surface through mussel-mimetic catechol chemistry, carrying biofunctional ligands that construct an artificial cellular microenvironment on the surface. The figure was reconstituted from references 18 and 19.*

spectroscopy (XPS) analysis (Figure 3). Culturing human umbilical vein endothelial cells (HUVECs) on REDV-PA/DOPA-PA coated stainless steel selectively favoured and enhanced endothelial cell activity. For example, compared to non-functionalized (i.e. bare) stainless steel surface, the number of endothelial cells adhered to REDV-PA/DOPA-PA nanofibers was dramatically increased *in vitro* (Figure 4). On the other hand, smooth muscle cell adhesion remained comparable on both groups. In addition, endothelial cells could spread and build polymerized actin filaments *in vitro* on REDV-PA/DOPA-PA nanofibers within 2 h, even in the absence of serum. These results suggest that REDV-PA/DOPA-PA nanofibrous coating can address major obstacles of stent technology by combining reliable, biocompatible surface coating technology with integrin-mediated bioactivity in order to promote selective endothelialization on a stainless steel surface.

Using a similar approach, we developed another hybrid biofunctional coating for orthopedic and dental implants in order to enhance the osseointegration rate, which was assessed by early osteogenic activity of bone-generating cells on TiAl6V4 surfaces. For this purpose, a KRSR-PA/DOPA-PA nanofiber-based coating was developed (Table 2). KRSR ligand binds to transmembrane proteoglycans and promotes adhesion of osteoblasts, while inhibiting the adhesion of fibroblasts.[77] XPS analysis confirmed functionalization of TiAl6V4 with KRSR-PA/DOPA-PA nanofibers (Figure 3). Statistically significant osteoblast adhesion, enhanced alkaline phosphatase (ALP) activity, and higher calcium ion deposition on KRSR-PA/DOPA-PA coated TiAl6V4 compared to bare TiAl6V4 surface were considered to be evidence of enhanced osteogenic activity (Figure 5).

This strategy can be extended to create a wide range of biological activities on any desired surface. Through altering the biofunctional group on *XXX*-PA/DOPA-PA model system, the bioactivity of nanofibers can be completely transformed. Potential applications of this adhesive model include cell-selective implant coatings, tissue adhesives, and biofunctional sealants.

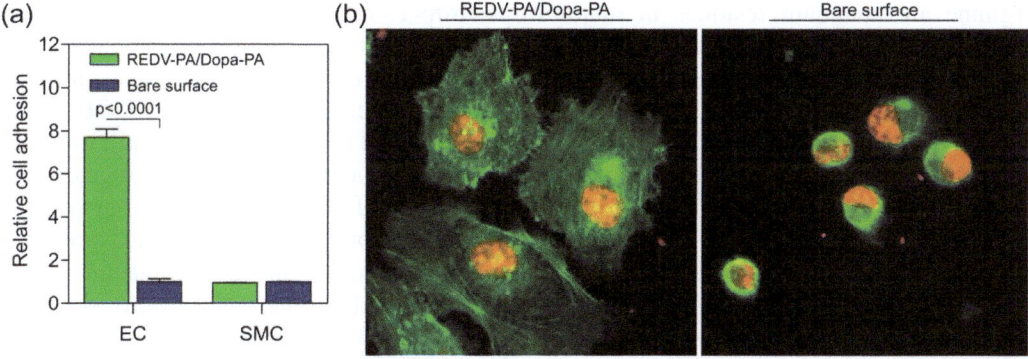

Figure 4   *(a) Relative adhesion of endothelial (EC) and smooth muscle cells (SMC) on REDV-PA/DOPA-PA nanofiber functionalized stainless steel surface compared to bare surface. (b) Initial spreading of endothelial cells at 2 h. Green regions indicate filamentous actin fibers whereas red regions indicate cell nuclei. The figure was reconstituted from reference 19.*

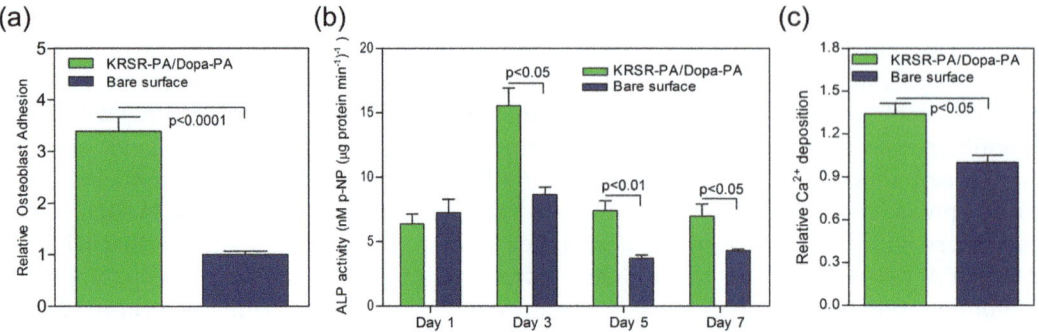

Figure 5   *(a) Relative adhesion of osteoblasts on KRSR-PA/DOPA-PA nanofiber functionalized TiAl6V4 substrate compared to bare surface. (b) Alkaline phophatase activity of osteoblasts. (c) Calcium deposition onto TiAl6V4 substrate on day 14. The figure was reconstituted from reference 18.*

5 FUTURE PERSPECTIVES

As a reductionist approach, DOPA or catechol-modification has proven very practical for functionalization of both bulk and nanoparticle surfaces. However, these biomimetic adhesives are far from reaching the exceptional adhesive and mechanical performance of the original glue. In mussels, the hierarchical organization of a number of different types of proteins, including collagens with silk- and elastin-like domains, mfps with varying DOPA and other amino acid contents, determine the overall performance of the final glue. Consequently, it is imperative to elucidate the functional roles of other chemical groups and conditions under which DOPA adhesion is the optimal. For instance, lysine and arginine are abundant residues in major mfps, mfp-1, mfp-3, and mfp-5, imparting a cationic character to the mussel glue.[16,27,29,45,78] This is not unique to mussels. The sandcastle worm, *Phragmatopoma californica*, lives in a protective tubular shelter, formed by gluing inorganic particles together with a special glue, which is rich in DOPA and lysine.[79-80] Although the particular role of having lysine and DOPA together is still unclear, the commonality of this system among natural underwater adhesion systems indicates that an exclusive interaction and/or cooperation between DOPA and lysine may provide a universal solution for surviving the challenging conditions of the seashores. For this reason, many synthetic materials mimicking the mussel adhesion mechanism have focused on utilizing DOPA and Lys (or its positively charged amine side chain) residues together.[19,23,27,36,45,81-82] On the other hand, high positive charge density was shown to be detrimental for surface adhesion.[15] Consequently, designing a novel material should take positive charge density into account for achieving the most efficient adhesion strength.

Although oxidation of DOPA into quinone is known to be detrimental for surface adhesion,[25] a certain degree of oxidation is necessary for mfp-1 adsorption onto surfaces.[15] At pH 4-5, mfp-1 chains are prevented from interacting in solution due to diminished DOPA oxidation and high positive charge density.[83] Under these conditions, adsorption of mfp-1 on surfaces forms a monolayer according to the Langmuir equation.[83] Haemers et al. examined mfp-1 adsorption to quartz at higher pH (6-8) under conditions where DOPA oxidation was allowed to some extent prior to protein adsorption. At low mfp-1 concentrations, intramolecular cross-linking dominated the system, culminating in fewer surface contacts and deposition of a stiffer film during mfp-1 adsorption.[83] At higher mfp-1 concentrations, crosslinks were mostly intermolecular, and the adsorbed protein was attached to a number of surface contacts, resulting in deposition of additional layers.

Phosphate is another intriguing group that is found in the underwater adhesive of the sandcastle worm.[31] This group imparts a water-resistant character to sandcastle worm adhesive through interaction with amine, Mg^{2+} and Ca^{2+} cations forming a complex coacervate.[84] Increasing pH rapidly hardens this coacervate within seconds followed by slower covalent curing through catechol oxidation. Therefore, grafting phosphate residues in addition to lysine and DOPA with optimized density and backbone design could develop a new generation of biomimetic adhesives.

6 CONCLUSION

Biomimetic materials has emerged as a converging discipline to reconstitute adaptive characteristics of biological systems in synthetic platforms, applying these to solve structural and functional problems in engineering, materials science, and medicine. A high performance adhesive, stable in aqueous and saline environments but with surface versatility and environmental compatibility will find a broad range of applications in industry and medicine. Although natural underwater adhesives exhibit exceptional performance under highly abrasive conditions, the high cost of obtaining adhesives from their original natural sources promotes alternative biomimetic solutions. Because of the simplicity of catechol that allows easy grafting to synthetic materials and the versatility of substrates it could bind to, mussel-inspired surface functionalization has become a prevalent strategy. Due to its general biocompatibility and water-resistant character, the main application area of mussel-mimetic materials has been medical applications. Unique underwater adaptations of other aquatic organisms also offer potential opportunities and novel inspirations for the purpose of developing advanced functional materials. Improving their efficiency and conditions under which synthetic adhesives operate is a continuing venture to meet the ever-changing demands of industry. This process is highly analogous to the evolution of natural adhesives under the force of natural selection in order to meet the ever-changing conditions of the environment.

Acknowledgements

Authors acknowledge support of the Scientific and Technological Research Council of Turkey (TUBITAK) grant number 110M353, COMSTECH-TWAS grant, Marie Curie International Reintegration Grant. H.C. is supported by TUBITAK BIDEB PhD fellowship. M.O.G. acknowledges support from the Turkish Academy of Sciences Distinguished Young Scientist Award (TUBA GEBIP).

References

1   C. S. Wang and R. J. Stewart, *J. Exp. Biol.*, 2012, **215**, 351-361.
2   S. Kaur, G. M. Weerasekare and R. J. Stewart, *ACS Applied Mater. Interface.*, 2011, **3**, 941-944.
3   B. D. Winslow, H. Shao, R. J. Stewart and P. A. Tresco, *Biomaterials*, 2010, **31**, 9373-9381.
4   C. Sun, G. E. Fantner, J. Adams, P. K. Hansma and J. H. Waite, *J. Exp. Biol.*, 2007, **210**, 1481-1488.
5   J. H. Waite, *Integr. Comp. Biol.*, 2002, **42**, 1172-1180.

6   J. H. Waite and C. C. Broomell, *J. Exp. Biol.*, 2012, **215**, 873-883.

7   J. H. Waite and X. Qin, *Biochemistry*, 2001, **40**, 2887-2893.

8   J.-L. Jonker, J. von Byern, P. Flammang, W. Klepal and A. M. Power, *J. Morphology.*, 2012, **273**, 1377-1391.

9   D. K. Burden, D. E. Barlow, C. M. Spillmann, B. Orihuela, D. Rittschof, R. K. Everett and K. J. Wahl, *Langmuir*, 2012, **28**, 13364-13372.

10   I. Y. Phang, N. Aldred, A. S. Clare and G. J. Vancso, *J. R. Soc. Interface*, 2008, **5**, 397-402.

11   R. J. Stewart, T. C. Ransom and V. Hlady, *J. Polymer Sci. B: Polymer Phys.*, 2011, **49**, 757-771.

12   R. G. Richards, T. F. Moriarty, T. Miclau, R. T. McClellan and D. W. Grainger, *J Orthopaedic Trauma.*, **26**, DOI: 10.1097/BOT.0b013e31826e37a2.

13   G. Avila, K. Misch, P. Galindo-Moreno and H.-L. Wang, *Implant Dentistry*, 2009, **18**, 17-26

14   Y. Liu, J. P. Li, E. B. Hunziker and K. de Groot, *Phil. Trans. R. Soc. A*, 2006, **364**, 233-248.

15   N. Holten-Andersen and J. H. Waite, *Journal of Dental Research*, 2008, **87**, 701-709.

16   B. P. Lee, P. B. Messersmith, J. N. Israelachvili and J. H. Waite, *Ann. Rev. Mater. Res.*, 2011, **41**, 99-132.

17   J. H. Waite, N. H. Andersen, S. Jewhurst and C. Sun, *J. Adhesion.*, 2005, **81**, 297-317.

18   H. Ceylan, S. Kocabey, A. B. Tekinay and M. O. Guler, *Soft Matter*, 2012, **8**, 3929-3937.

19   H. Ceylan, A. B. Tekinay and M. O. Guler, *Biomaterials*, 2011, **32**, 8797-8805.

20   L. A. Burzio and J. H. Waite, *Protein Sci.*, 2001, **10**, 735-740.

21   L. A. Burzio and J. H. Waite, *Biochemistry*, 2000, **39**, 11147-11153.

22   N. Holten-Andersen, M. J. Harrington, H. Birkedal, B. P. Lee, P. B. Messersmith, K. Y. C. Lee and J. H. Waite, *Proc. Nat. Acad. Sci.*, 2011, **108**, 2651-2655.

23   H. Ceylan, M. Urel, T. S. Erkal, A. B. Tekinay, A. Dana and M. O. Guler, *Adv. Funct. Mater.*, 2012, DOI: 10.1002/adfm.201202291

24   D. S. Hwang, M. J. Harrington, Q. Lu, A. Masic, H. Zeng and J. H. Waite, *J. Mater. Chem.*, 2012, **22**, 15530-15533.

25   H. Lee, N. F. Scherer and P. B. Messersmith, *Proc. Nat. Acad. Sci.*, 2006, **103**, 12999-13003.

26   D. J. Crisp, G. Walker, G. A. Young and A. B. Yule, *J. Coll. Interface Sci.*, 1985, **104**, 40-50.

27   H. Lee, S. M. Dellatore, W. M. Miller and P. B. Messersmith, *Science*, 2007, **318**, 426-430.

28   M. J. Harrington, A. Masic, N. Holten-Andersen, J. H. Waite and P. Fratzl, *Science*, 2010, **328**, 216-220.

29   V. V. Papov, T. V. Diamond, K. Biemann and J. H. Waite, *J. Biol. Chem.*, 1995, **270**, 20183-20192.

30   H. J. Cha, D. S. Hwang and S. Lim, *Biotech. J.*, 2008, **3**, 631-638.

31   H. Zhao and J. H. Waite, *J. Biol. Chem.*, 2006, **281**, 26150-26158.

32   L. M. Hamming, X. W. Fan, P. B. Messersmith and L. C. Brinson, *Composites Sci. Tech.*, 2008, **68**, 2042-2048.

33   M. Rief, H. Clausen-Schaumann and H. E. Gaub, *Nat. Struct. Mol. Biol.*, 1999, **6**, 346-349.

34   J. Monahan and J. J. Wilker, *Langmuir*, 2004, **20**, 3724-3729.

35   H. Xu, J. Nishida, W. Ma, H. Wu, M. Kobayashi, H. Otsuka and A. Takahara, *ACS Macro Lett.*, 2012, **1**, 457-460.

36 H. Lee, J. Rho and P. B. Messersmith, *Adv. Mater.*, 2009, **21**, 431-434.

37 E. Carrington, *Integr. Comp. Biol.*, 2002, **42**, 846-852.

38 J. Gosline, M. Lillie, E. Carrington, P. Guerette, C. Ortlepp and K. Savage, *Phil. Trans. R. Soc. B*, 2002, **357**, 121-132.

39 J. H. Waite, H. C. Lichtenegger, G. D. Stucky and P. Hansma, *Biochemistry*, 2004, **43**, 7653-7662.

40 J. H. Waite, E. Vaccaro, C. Sun and J. M. Lucas, *Phil. Trans. R. Soc B,* 2002, **357**, 143-153.

41 J. H. Waite, *J, Biol, Chem.*, 1983, **258**, 2911-2915.

42 A. Avdeef, S. R. Sofen, T. L. Bregante and K. N. Raymond, *J. Amer. Chem. Soc.*, 1978, **100**, 5362-5370.

43 N. Holten-Andersen, G. E. Fantner, S. Hohlbauch, J. H. Waite and F. W. Zok, *Nat. Mater.*, 2007, **6**, 669-672.

44 K. Inoue, Y. Takeuchi, D. Miki and S. Odo, *J. Biol. Chem.*, 1995, **270**, 6698-6701.

45 J. D. White and J. J. Wilker, *Macromolecules*, 2011, **44**, 5085-5088.

46 G. A. Parks, *Chem. Rev.*, 1965, **65**, 177-198.

47 L. L. Schramm, K. Mannhardt and J. J. Novosad, *Colloids Surf.*, 1991, **55**, 309-331.

48 S. C. Warner and J. H. Waite, *Mar. Biol.*, 1999, **134**, 729-734.

49 H. Zhao and J. H. Waite, *Biochemistry*, 2006, **45**, 14223-14231.

50 J. Yu, W. Wei, E. Danner, R. K. Ashley, J. N. Israelachvili and J. H. Waite, *Nat. Chem. Biol.*, 2011, **7**, 588-590.

51 J. Yu, W. Wei, E. Danner, J. N. Israelachvili and J. H. Waite, *Adv. Mater.*, 2011, **23**, 2362-2366.

52 H. Silverman and F. Roberto, *Mar. Biotech.*, 2007, **9**, 661-681.

53 D. S. Hwang, H. J. Yoo, J. H. Jun, W. K. Moon and H. J. Cha, *Appl. Environ. Microbiol.*, 2004, **70**, 3352-3359.

54 D. R. Filpula, S.-M. Lee, R. P. Link, S. L. Strausberg and R. L. Strausberg, *Biotech. Prog.*, 1990, **6**, 171-177.

55 A. J. Salerno and I. Goldberg, *Appl. Microbiol. Biotech.,* 1993, **39**, 221-226.

56 M. Kitamura, K. Kawakami, N. Nakamura, K. Tsumoto, H. Uchiyama, Y. Ueda, I. Kumagai and T. Nakaya, *J. Polymer Sci. A*, 1999, **37**, 729-736.

57 B.-H. Choi, Y. S. Choi, D. G. Kang, B. J. Kim, Y. H. Song and H. J. Cha, *Biomaterials*, 2010, **31**, 8980-8988.

58 J. L. Dalsin and P. B. Messersmith, *Materials Today*, 2005, **8**, 38-46.

59 J. L. Dalsin, L. Lin, S. Tosatti, J. Vörös, M. Textor and P. B. Messersmith, *Langmuir*, 2004, **21**, 640-646.

60 J.-Y. Wach, B. Malisova, S. Bonazzi, S. Tosatti, M. Textor, S. Zürcher and K. Gademann, *Chemistry − A European Journal*, 2008, **14**, 10579-10584.

61 J. H. Ryu, Y. Lee, W. H. Kong, T. G. Kim, T. G. Park and H. Lee, *Biomacromol.*, 2011, **12**, 2653-2659.

62 H. Lee, B. P. Lee and P. B. Messersmith, *Nature*, 2007, **448**, 338-341.

63 D. Ling, W. Park, Y. I. Park, N. Lee, F. Li, C. Song, S.-G. Yang, S. H. Choi, K. Na and T. Hyeon, *Angewandte Chemie Intern. Ed.*, 2011, **50**, 11360-11365.

64 E. Kim, Y. Liu, X.-W. Shi, X. Yang, W. E. Bentley and G. F. Payne, *Adv. Funct. Mater.*, 2010, **20**, 2683-2694.

65 S. Hong, Y. S. Na, S. Choi, I. T. Song, W. Y. Kim and H. Lee, *Adv. Funct. Mater.*, 2012, **22**, 4711-4717.

66 J. Ryu, S. H. Ku, H. Lee and C. B. Park, *Adv. Func. Mater.*, 2010, **20**, 2132-2139.

67 H. Wei, N. Insin, J. Lee, H.-S. Han, J. M. Cordero, W. Liu and M. G. Bawendi, *Nano Lett.*, 2011, **12**, 22-25.

68 S. M. Kang, N. S. Hwang, J. Yeom, S. Y. Park, P. B. Messersmith, I. S. Choi, R. Langer, D. G. Anderson and H. Lee, *Adv. Funct. Mater.*, 2012, **22**, 2949-2955.

69 H. Cui, M. J. Webber and S. I. Stupp, *Peptide Sci.*, 2010, **94**, 1-18.

70 T. Dvir, B. P. Timko, D. S. Kohane and R. Langer, *Nat. Nano.*, 2011, **6**, 13-22.

71 T. D. Sargeant, S. M. Oppenheimer, D. C. Dunand and S. I. Stupp, *J. Tissue Eng. Regenerative Med.*, 2008, **2**, 455-462.

72 G. A. Silva, C. Czeisler, K. L. Niece, E. Beniash, D. A. Harrington, J. A. Kessler and S. I. Stupp, *Science*, 2004, **303**, 1352-1355.

73 S. Zhang, *Nat. Biotech.*, 2003, **21**, 1171-1178.

74 M. O. Guler and S. I. Stupp, *J. Amer. Chem. Soc.*, 2007, **129**, 12082-12083.

75 M. O. Guler, S. Soukasene, J. F. Hulvat and S. I. Stupp, *Nano Lett.*, 2004, **5**, 249-252.

76 J. A. Hubbell, S. P. Massia, N. P. Desai and P. D. Drumheller, *Nat. Biotech.*, 1991, **9**, 568-572.

77 K. C. Dee, T. T. Andersen and R. Bizios, *J. Biomed. Mater. Res.*, 1998, **40**, 371-377.

78 J. H. Waite, *J. Biol. Chem.*, 1983, **258**, 2911-2915.

79 C. E. Brubaker and P. B. Messersmith, *Langmuir*, 2012, **28**, 2200-2205.

80 J. H. Waite, R. A. Jensen and D. E. Morse, *Biochemistry*, 1992, **31**, 5733-5738.

81 M. Yu and T. J. Deming, *Macromolecules*, 1998, 31, 4739-4745.

82 P. Podsiadlo, Z. Liu, D. Paterson, P. B. Messersmith and N. A. Kotov, *Adv. Mater*, 2007, **19**, 949-955.

83 S. Haemers, M. C. van der Leeden, G. J. M. Koper and G. Frens, *Langmuir*, 2002, **18**, 4903-4907.

84 H. Shao and R. J. Stewart, *Adv. Mater.*, 2010, **22**, 729-733.

CHARACTERIZATION OF BIOMIMETIC ADHESIVES FROM THE RED ALGA *GRACILARIA CONFERTA* FOR BIOMEDICAL APPLICATIONS

S. Dimartino*[1], I. Lir[2], M. Haber[2] and R. Azhari[3]

[1] Biomolecular Interaction Centre, University of Canterbury, Christchurch, New Zealand
[2] Biota Ltd., Or-Akiva, Israel
[3] Department of Biotechnology Engineering, ORT Braude College, Karmiel, Israel
*simone.dimartino@canterbury.ac.nz

1   INTRODUCTION

The production of adhesives to be used in wet environments represents a crucial issue for the marine and food process industries, and finds a variety of biomedical applications such as in tissue engineering, medical implants and tissue adhesives.[1-5] Recent trends in material science demonstrate the potential of underwater adhesives inspired by marine organisms such as mussels,[6,7] barnacles[8] and sandcastle worms,[9] all capable of secreting adhesives that bind strongly to virtually all inorganic and organic surfaces in aqueous environments.[10]

Marine macroalgae represent an additional class of organisms able to produce strong water resistant adhesives. They evolved remarkable underwater adhesives endowed with rapid and permanent attachment to withstand the strong mechanical stresses and pullout forces associated with wave swept shores and tidal currents.[11-13] Marine algae are able to cling to a wide range of substrates such as rocks, glass and metals under extreme conditions of temperature, salinity and turbulence, indicating their adhesive capability.[14-16] In addition, they are highly resistant to mechanical removal and also tolerate chemical preventive agents such as toxic biocides.[17] The responsible bioadhesives have extraordinarily high cohesive and adhesive strength to solid surfaces, enabling the organisms to remain attached under conditions of tension comparable to those found in a surgical environment.[18,19] These qualities indicate a promising avenue in the development of effective tissue adhesives for medical use, intended to replace painful traditional wound closure methods or applied as bioadhesive matrices in controlled release systems for skin and buccal delivery of active agents.[19]

Macroalgae are exploited worldwide as commercial sources for a variety of phycocolloids and other biochemicals that are utilized in products such as thickeners, emulsifiers, stabilizers and adhesives which are extensively used in the pharmaceutical,[19] cosmetic, food[20] and biomedical industries.[21] Algal polysaccharides, especially alginates, are mainly applied in wound care and the surgical arena, as means for treatment of acute and chronic wounds. Alginate hydrogels and fibrous products such as wound dressings and scaffolds for tissue engineering enable improved wound healing and tissue regeneration.

Different research groups have contributed to shed light on the secretion mechanism,[22-27] biochemical composition[28-31] and role of the constituents in the biological glue from macroalgae.[32-35] In general, algal bioadhesives consist of a complex mixture of macromolecules, mainly proteins, polysaccharides, glycoproteins and polyphenols.[36,37] In the specific case of red algae, sulphated polysaccharides are the main constituent of the bioadhesives,[38,39] but a number of other different polysaccharides based on α-D mannose, α-D glucose, β-D galactose, N-acetyl-glucosamine and N-acetyl galactosamine have been reported[40,41] as well as the presence of a proteinaceous component.[42,43] However, reliable procedures for the extraction and purification of the adhesive components are complex, making it a challenge to achieve effective isolation and precise biochemical characterization of the bioadhesive.[44] The exact composition of algal bioadhesives and the associated curing process are largely unknown to date; in addition, the little amount of adhesive material present in the holdfast limits direct analysis on living samples.

An attempt to extract and isolate adhesive polymers from whole red-algae *Gracilaria conferta* biomass is described in this chapter. The algal adhesive was obtained following the method described by Haber[45] to extract bioadhesives from algal holdfasts, assuming that isolated polymers would have homology to substrate bioadhesives. The biochemical characterization of the biopolymers extracted is also presented. The extracts were considered in two different biomedical applications, namely to produce films for burn dressing and wound closure, and as a biocompatible adhesive. Finally, this chapter includes an overview on algal products currently investigated in the biomaterial arena, highlighting recent developments and applications in the biomedical industry.

2   EXPERIMENTAL

2.1 Materials

G. conferta plants were obtained from Rahan Seaweed Ltd. (Rosh-Hanikra, Israel), which cultivates this red algae species at the eastern Mediterranean coast of northern Israel.

Glycerol and ethyl-3-(3-dimethylaminopropyl)carbodiimide were purchased from Sigma Aldrich and used to plasticize and crosslink *G. conferta* extracts, respectively. Gelatin derived from porcine skin collagen with 185 Bloom gel strength was purchased from Sigma Aldrich and used for film preparation. All other chemicals used throughout this work were also purchased from Sigma Aldrich.

Commercial polymer films for skin wound dressings were evaluated as benchmark mechanical and adhesion tests: Omiderm™ (ITG Laboratories Inc., Redwood City, CA, USA), Tegaderm™ (3M Inc., St Paul, Minnesota, USA) and Rexam RXM 1101 (Rexam, London, UK). A commercial cyanoacrylate based glue (Super Glue, Loctide Ltd., Westlake, Ohio, USA), was used as positive control.

Cellulose acetate MWCO 3,500 (Spectrum Lab. Inc., Rancho Dominguez, CA, USA) and Mylar™ polyester (DuPont, Hopewell, VA, USA) were used as reference materials in the adhesion tests.

2.2 Extraction and Purification of Biopolymers from *G. conferta* Plants

Extraction and purification of algal biopolymers from *G. conferta* were performed following a procedure similar to the one described by Haber.[45] The bio-glue from macroalgae has been described as intrinsically insoluble where the carbohydrate and protein molecules involved form a closely cross-linked network.[16,36] For this reason the protocol is based on the extraction of biopolymer aggregates followed by hydrolysis to allow their recovery in solution and ease further handling. The insoluble matter present after hydrolysis mainly contains hydrophobic components which are less likely to be involved in the adhesive formulation.

In detail, 17 kg of freshly harvested whole mature *G. conferta* plants were kept at 4°C for two days. The algal material was ground for 10 minutes in a Fitz Mill Model D equipped with a ½ inch sieve (Fitzpatrick, Elmhurst, IL, USA). 5 L of cold water were then added to the grinded algal mass and the resulting mixture was milled at 10°C for 15 minutes using Fryma Knives (Fryma AG, Rheinfelden, Switzerland) and subsequently homogenized for 45 minutes using Gaulin 15M homogeniser (Gaulin, Delavan, WI, USA) at a pressure of 500 atm resulting in cell breakage. The obtained homogenate, having a pH of 6.3, was centrifuged at 4000 rpm for 45 minutes at 25°C using an Escher Wyss H-400 centrifuge (Escher-Wyss, Houston, TX, USA). The supernatant containing water soluble algal materials was discharged, while 8 kg of the pellet were collected for further extraction of cross-linked and non-soluble algal biopolymers. 11 L of cold water containing 0.1% v/v Tween-80 and 0.15% v/v of a 85% phosphoric acid solution were added to the precipitate to allow partial hydrolysis of the cross-link bonds in the adhesive molecules. The mixture was homogenized again at 25°C for 25 minutes and then centrifuged at 5000 g for 10 minutes at 25°C using a Westphalia centrifuge LWA-205 (Westphalia, Niederahr, Germany). In this step the precipitate containing cell debris and highly hydrophobic compounds was discharged, while the supernatant, having a pH of 3.5, was mixed with 2 volumes of ethanol to precipitate the algal biopolymers overnight at -20°C. The precipitate was finally filtered through a Buchner filter funnel and stored at -25°C. The final mass of extract was 1530 g, corresponding to an overall yield of 9%.

2.3 Biochemical Characterization of *G. conferta* Algal Extract

Characterization of the algal extract was accomplished by carrying out a number of different biochemical tests as described below.

Protein content in the extract was determined applying the Lowry method using bovine serum albumin as a standard.[46,47] Amino acid composition was investigated by acid hydrolysis followed by HPLC amino acid analysis following the methods described by Moore.[48,49] Elemental analysis was obtained by Inductively Coupled Plasma Mass Spectrometry.[50]

Assays employed to quantify charged sugars of the polysaccharide were sulfate content, uronic acid and hexoseamines. These assays were performed after hydrolysis of the samples with trifluoroacetic acid as described by Ray.[51] Uronic acid content was assessed using the carbazole reaction test with glucuronic acid as standard,[52] while hexoseamines were determined by the Elson-Morgan assay using glucosamine as standard.[53] Sulfate content was measured by applying the rheodizonate method using sulfuric acid as a standard.[54]

2.4 Preparation of Thin Films from *G. conferta* Extract

Algal films were prepared following the casting technique, as described by Mark *et al.*[55] Briefly, 3 g of algal extract were homogenized with 30 ml of distilled water in a glass-Teflon Potter-Elvehjem homogenizer at 25°C for 3 minutes. The homogenate was cast onto a square polystyrene dish (10 x 10 cm, Bibby Sterilin, Newport, UK) and dried at 25°C for 24 hours followed by vacuum drying until constant weight. Films manufactured with this method had a thickness of 30 ± 5 μm.

Two chemical treatments, namely plasticization and cross-linking, were also considered on the algal extract to modify and tune its mechanical and adhesive properties.

Plasticized films were produced following the same protocol described for the unmodified films but with the addition of 0.03 g glycerol to the algal extract/water mixture prior to homogenization. The thickness of the plasticized films was 30 ± 5 μm, practically the same as for the unmodified ones.

Cross-linking was performed on a 100 cm^2 piece of dry film (unmodified or plasticized) soaked in 100 ml of 2 mg/ml EDC solution in acetone-water (9:1) at 25°C for 4 hr. The samples were then carefully washed with fresh acetone-water mixture to remove excess of EDC. The film was air dried at 25°C and then finally vacuum dried until constant weight. The thickness of the cross-linked films was practically the same as plasticized and unmodified ones, indicating that no major swelling or shrinking with respect to the original film occurred during EDC treatment.

2.5   Mechanical Tests on Algal Films

Tensile properties of the films were evaluated according to "Standard Test Method for Tensile Properties of Thin Plastic Sheeting" described in ASTM standard D882-10.[56] A Lloyd Universal Materials Testing Machine LRX-plus (Lloyd Instruments, West Sussex, UK) equipped with a 5 kN load cell and self tightening grips TG-20 to hold the specimens was used for the measurements. The instrument was controlled by a Lloyd Nexygen 4.0 software package supporting the library of international standards for mechanical testing. Dry films were cut into 100 x 10 mm strips, clamped to the equipment and stretched uniaxially at a constant grip separation rate of 80 mm/min. Stress–strain curves were recorded for five samples of each type of film produced. Films were uniaxially stretched until the break point was reached. All tests were performed at a temperature of 25°C and relative humidity of 50%.

2.6   Adhesion Tests

The adhesive capacity of the algal extract was evaluated by performing lap shear tests accordingly to the "Standard Test Method for Tensile Properties of Adhesive Bonds" reported in ASTM standard D897-08.[57] As the substrate plays a major role in the overall adhesion strength and final properties of the joint, several adherends differing in their surface properties were considered in the present study.

Cellulose acetate and polyester Mylar were chosen as example of hydrophilic and hydrophobic surfaces, respectively. The cellulose acetate and polyester supports used in the adhesion tests had a thickness of 100 μm and were cut into 100 x 10 mm strips.

In addition to cellulose acetate and Mylar, gelatin films were considered as a model material to mimic a biological tissue. Gelatin films were prepared using 185 Bloom gelatin derived from porcine skin collagen. Briefly, 5 g of gelatin were dissolved in 95 ml warm distilled water (55°C). The solution was then cast on a polystyrene dish and dried at 25°C for 24 hours. The resulting film, with a thickness of 100 ± 10 μm, was cut into 100 x 10 mm strips.

To prepare the adhesive joints, 15 μL of the algal extract were spread onto the ends of two substrate strips (20 mm length) and the two surfaces were joined together. To secure the attachment, the two strips were pressed to each other by rolling a 1 kg weight, twice, over the juncture and subsequently applying a force of 1 kg for 5 min. The specimens were then immediately stretched uniaxially at a constant rate of 80 mm/min at 25°C and relative humidity of 50% up to joint failure.

In order to assess the adhesive performances of the algal extracts, Super Glue and distilled water were used as positive and negative control, respectively. Ten independent measurements were performed for each type of algal extract investigated.

2.7 Biocompatibility of *G. conferta* Extract

In vitro cytotoxicity was evaluated using the agarose overlay method towards L-929 mouse fibroblast cells according to the "Biological evaluation of medical devices - Part 5: Tests for in vitro cytotoxicity" ISO 10993-5 standard.[58] *In vivo* acute intracutaneous reactivity test was conducted in three rabbits according to the "Biological evaluation of medical devices - Part 10: Tests for irritation and skin sensitization" ISO 10993-10 standard.[59] The rabbit skin was carefully monitored during 72 hours after injection. *In vivo* acute systemic toxicity test was conducted in five mice according to the "Biological evaluation of medical devices - Part 11: Tests for systemic toxicity" ISO 10993-11 standard[60] using an intravenous or intraperitoneal systemic injection.

3 RESULTS AND DISCUSSION

3.1 Biochemical Composition of *G. conferta* Algal Extract

Results of the biochemical tests revealed that the main ingredients of *G. conferta* extract were glycoproteins and polysaccharides, with weight percentages of 20% and 70%, respectively. As summarized in Table 1, the protein fraction possessed a high ratio of the negatively charged glutamic and aspartic acids. Consistent with the abundance of aspartate, glycine and arginine residues, RGD and RGD-like adhesion recognition sequences are likely to be present. RGD like sequences have been reported in the vitronectin-like glycoproteins discovered in other red macroalgae species and are thought to be involved in cell-cell and cell-surface adhesion.[33,61] Interestingly, neither of the two putative adhesive amino acids found in mussel adhesive proteins, namely DOPA (3,4-dihydroxyphenylalanine) and hydroxyproline,[62,63] were detected in the algal extract. Presence of these two post translational modified amino acids has never been reported for macroalgae, and although traces may be present, the role of these two moieties in attachment must be very limited and cannot account for the strong and durable adhesion of macroalgae. This observation indicates that the adhesion strategy undertaken by marine macroalgae is based on different adhesive blocks than DOPA or hydroxyproline.

Carbohydrate analysis pointed out a relatively rich content of negative polysaccharides, with hexuronic acid and sulphated sugars accounting for the 4.6 % and 0.5 % molar fraction of the extract. The high content of sulphur observed in the elemental analysis

Table 1 *Main amino acids in the protein fraction of the algal extract*

Amino acid	Abbreviation	Mol. frac. %
Glutamine	E	15.0
Aspartate	D	13.0
Glycine	G	9.0
Alanine	A	7.6
Arginine	R	6.0
Cysteine	C	1.2

(Table 2) suggests that sulphated polysaccharides may be present in the extract in much greater amounts. In fact it is worth considering that during extraction and sample preparation the sulphated groups could be hydrolysed from the sugar ring, with a consequent underestimation of sulphated polysaccharides content. Sulphated polysaccharides such as agar and carrageenan are abundant in the cell wall of red algae[38] and are thought to possess a major role in algal adhesion.[40,41,64] The abundance of hexuronic acids is associated to alginates which also play an important role in algal adhesion.

The main feature of the two classes of polysaccharides is in their negative charge, which confers to the bio-glue both adhesive and cohesive functions. Adhesion is in fact mainly achieved through long and short range electrostatic interactions with the substrate surface,[65] while cohesion is established thanks to coordination bonds mediated by metal cations entrapped in the adhesive gel.[66] Occurrence of such cross-linked structures in the algal extract from *G. conferta* is supported by the relatively high content of Ca and Mg ions found (Table 2).

Hexoseamines were also present in the extract at a molar concentration of 4.5%. Contrary to alginates and sulphated polysaccharides, hexoseamines have a positive or neutral charge and are ascribable to the presence of chitosan-like chains in the algal adhesive. The intimate association between alginate and chitosan chains has structural significance in the formation of a solid adhesive scaffold that improves the cohesive characteristics of the algal extract.[67]

As reported in Table 2, content of Fe, Cu and Zn ions in the algal extract is very limited and not comparable to the high levels detected in mussel proteins. Transition metals ions are known to play a major role in the formation of the structure of the mussel adhesive pad.[68] This observation further indicates that the adhesive mechanisms followed by macroalgae and mussels differ in their intimate nature, as also confirmed by Vreeland and coworkers.[16]

Table 2   *Content of metal and other trace elements in the dry algal extract*

Element	Mass. frac. %
S	0.77
P	0.34
Na	0.34
Ca	0.19
Mg	0.17
K	0.12
Si	0.05
Fe	0.03
Cu	0.01
Zn	0.01
Al	0.01
Cd	<0.0001
Co	<0.0001
Hg	<0.0001
Pb	<0.0001

3.2 Mechanical Properties of *G. conferta* Algal Films

In order to improve and/or adjust the mechanical and surface properties of the algal films, two different chemical modifications were employed: plasticization with glycerol and/or chemical cross-linking with ethyl-3-(3-dimethylaminopropyl)carbodiimide (EDC). As reported by Avella for glycerol-plasticized alginates, addition of glycerol modifies the microstructure of the polymer thus more flexible and elastic material could be produced.[69] The synthetic cross-linker EDC was chosen because it catalyzes the formation of amide bonds between amine and carboxylic groups in an identical manner to the natural cross-linking enzyme transglutaminase. Accordingly to the cross-linking mechanism suggested by Choi,[70] carbodiimides activate the carboxylic groups of glutamic and aspartic acid residues, favouring their reaction with the amino groups of other polypeptide chains or glucosamine moieties in the polysaccharide fraction of the extract. In our approach, the carboxylic groups of both glycoprotein and polysaccharide fractions of the algal extract were activated by EDC, and then the activated carboxylic groups were expected to react with available amine groups. EDC is extensively used for cross-linking of biomaterials such as collagen, gelatin and others.[71,72] In addition, EDC is water soluble, so excess reagent and cross-linking byproducts are easily removed by simple water wash. It is expected that EDC treatment improves the strength and stability of the algal derived materials.

Figure 1 shows some typical stress-strain curves obtained with the algal films investigated; tensile strength and maximum elongation at break were derived from the stress-strain curves and are summarized in Table 3. To allow a better comparison, Table 3 also reports the data of other natural films and of commercial surgical products manufactured from synthetic polymers and clinically used for skin wound closure or burn healing.

The stress-strain curve for the non modified extract has a fairly linear trend, indicating that deformation on the film is associated to a truly elastic behaviour (Figure 1, curve A).

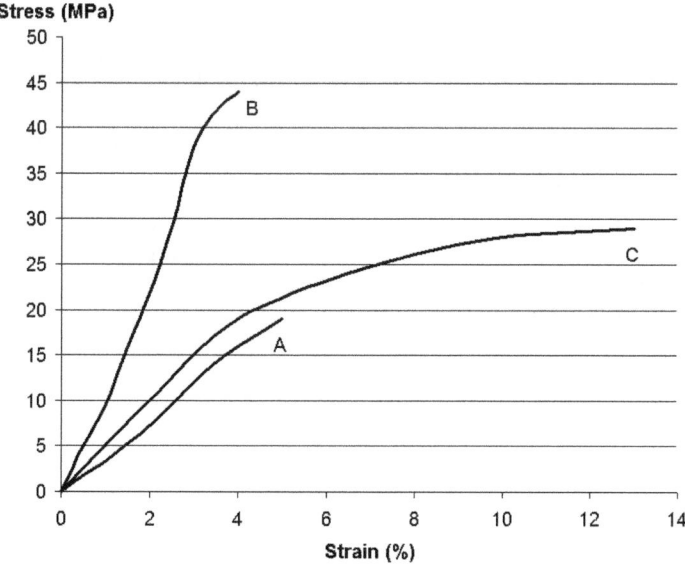

Figure 1 *Typical stress-strain curves obtained for algal films produced from unmodified extract (curve A), cross-linked film (curve B) and plasticized and cross-linked film (curve C)*

The unmodified film is very brittle, with no apparent plastic deformation until breakage. Treatment with glycerol enhanced the elastic behaviour of the film to the detriment of the maximum load borne, but did not introduce plastic behaviour to the film (data not shown). EDC cross-linking produced a film with the characteristic behaviour of a strong and stiff but not ductile material. This expected result confirms that covalent bonds have been created between the polysaccharide chains of the adhesive. Such bonds make the film stronger and introduce a small component of plastic deformation, particularly visible in the final curved part of the stress-strain curve, which is probably associated with changes in the mutual position of the carbohydrate fibres and rearrangement of the polymeric network (Figure 1, curve B). Combined plasticization and cross-linking of the algal films produced a more ductile material which could withstand higher elongations (Figure 1, curve C). Yet, the elastic properties of these treated films and films produced from the native algal extract were very similar. Interestingly, the transition from elastic (linear region) to plastic deformation (concave downward region) occurs at around 3% strain for both kinds of cross-linked materials, hence EDC based treatments operate similar changes in the structure of the film irrespective of glycerol treatment.

Compared to the natural plant protein films, the algal extracts have a much higher tensile strength. However, the plant protein films can sustain elongations larger than 100%, a performance not matched by any of the algal films produced which appear to be much more brittle (Table 3).

The film from unmodified algal extract has relatively poor properties compared to the commercial medical materials, having the same or even lower tensile strength but with limited ductility (Table 3). On the other hand, cross-linked films had a tensile strength comparable to or even higher than the commercial medical films. However, the cross-linked film had a very limited maximum elongation, similar to the unmodified film, still not satisfactory with respect to the commercial biomedical film. By combining plasticization and cross-linking a film with excellent mechanical properties, namely 30 MPa tensile strength and 12% maximum elongation, was obtained. These properties are comparable to the ones reported for the commercial materials. It is worth noticing that the film modification techniques proposed in this work enable regulation of the mechanical properties of the algal film, and therefore to produce films with properties fine tuned to match the requirements needed by specific biomedical applications.

Table 3   *Mechanical properties of* G. conferta *extract films and of natural and commercial films*

Film source	Film type	Tensile strength (MPa)	Max elongation (%)
Algal extract films	Unmodified algal extract	20 ± 3	4 ± 1
	Plasticized	10 ± 2	44 ± 2
	Cross-linked	42 ± 7	3 ± 1
	Plasticized and cross-linked	30 ± 3	12 ± 2
Plant protein films	Pea film[73]	3 ± 1	100 ± 17
	Soy film[73]	8 ± 0.2	300 ± 3
	Sunflower film[73]	4 ± 0.1	215 ± 14
Commercial skin burn dressings	Omiderm™	63 ± 2	11 ± 1
	Omiderm™ meshed	45 ± 2	7 ± 1
	Tegaderm™	22 ± 1	660 ± 30

3.3 Adhesive Capacity of *G. conferta* Extract

The adhesive capacity of *G. conferta* extracts was evaluated on a number of substrates with different surface characteristics. In fact, adhesion strength not only depends on the glue used but also on the specific adherend considered. In particular, the hydrophobic or hydrophilic characteristics of the substrate play a major role in adhesion and can drastically change the strength of the joint. Cellulose acetate was chosen as an example of a hydrophilic substrate possessing high wettability by aqueous media. This material is particularly interesting as it was previously considered as model substrate in other tests on bioadhesives.[74] A synthetic polyester, Mylar, i.e. a copolymer of ethylene glycol and dimethyl terephtalate, was used as a reference for hydrophobic surfaces. In addition, gelatin films were used as model substrate to mimic a living tissue.

Lap shear tests were carried out immediately after two strips of material were glued together. The value of the maximum stress withstood by the different joints tested is reported in Table 4. In addition to the native and treated algal extracts, experiments with water and a cyanoacrylate-based glue were carried out as negative and positive controls, respectively.

Bonding strength of the joint was found to be dependent on the type of adhesive, nature and surface characteristics of the substrate and the interaction between the adhesive and the substrate. The adhesive derived from *G. conferta* plants was capable of adhering to both hydrophobic and hydrophilic surfaces, however higher adhesion was observed in the case of hydrophilic adherends such as cellulose acetate and gelatin. In particular, *G. conferta* extract was able to securely bind hydrophilic materials with adhesive strength of up to 0.7 MPa, performance comparable to the cyanoacrylate-based glue. This surprising result indicates that the algal extract derived from G. *conferta* possesses strong adhesive properties already in its unmodified form.

When considering the hydrophobic surface, no difference in adhesive strength was observed among the different algal extracts, indicating that also in this case plasticization and/or cross-linking did not introduce a large variation in the binding properties. On hydrophobic surfaces, adhesion strength measured with the algal biopolymers was much weaker than the cyanoacrylate based glue. However this observation does not represent a limitation for biomedical applications of the algal extract from *G. conferta* as living tissues usually have a marked hydrophilic behaviour.

Table 4 *Results of lap-shear tests*

Adhesive	Lap-shear maximum strength (MPa)		
	Cellulose-Acetate MWCO 3500	*Gelatin*	*Mylar™*
Unmodified algal extract	0.54±0.10	0.70±0.10	0.37±0.03
Plasticized	0.47±0.12	0.54±0.08	0.35±0.02
Plasticized and cross-linked	0.56±0.04	0.50±0.01	0.38±0.07
Super Glue (positive control)	0.63±0.05	0.60±0.02	0.72±0.05
Distilled water (negative control)	0.20±0.05	0.25±0.05	0

3.4   Biocompatibility of *G. conferta* Extract

Extensive biocompatibility assays are paramount to demonstrate whether a new biomaterial has toxic or harmful effects in living tissues. The number and type of assays required is related to the composition of the products as well as to their intended use. Both *in vitro* and *in vivo* biocompatibility tests have been performed with *G. conferta* extracts. *In vitro* cytotoxicity tests showed no evidence of cell lyses or toxicity of mouse fibroblast cells after contact with the algal biopolymer. In order to determine if *G. conferta* extract would cause local dermal irritations, *in vivo* acute intracutaneous reactivity tests were conducted in rabbits. No evidence of irritation or toxicity was detected on the rabbit skin during the 72 h following injection. *In vivo* acute systemic toxicity tests performed on mice showed no mortality or evidence of systemic toxicity during the 72 hours following the intravenous or intraperitoneal injections. Although biocompatibility analysis must be further investigated and the absence of harmful effects associated to the algal extract must be supported by additional experiments, the preliminary biocompatibility characterization here presented shows promise for the use of the algal extract from *G. conferta* in the formulation of adhesives for biomedical applications.

4   BIOMEDICAL APPLICATIONS: AN OVERVIEW

Macro algae adhere to substrates in wet and stressful environments which closely resemble the conditions encountered in the human body.[18,19] This observation indicates that algal bioadhesives and related biomimetic approaches constitute a promising avenue towards the development of biocompatible adhesives. It is very likely that use of algal derived products would not be restricted to tissue adhesives, but it can be extensively used in a number of different applications in the biomedical industry. The review presented below provides insights to other relevant additional uses of algal or algal-inspired products in medicine.

The use of hemostats, sealants, and adhesives is rapidly changing and continues to gain significance in medical practice.[75] It appears impossible to develop an adhesive suitable for all kinds of living tissues because they present different functions, physico-chemical characteristics, rates of regeneration, types and amount of enzyme content, etc. Therefore, it appears necessary to synthesise and investigate several tailor made medical adhesives adjusting the product to the requirements of the intended application.[76]

Use of algal biopolymers for the production of biomedical adhesives has several advantages. First, as compared to human and animal derived biopolymers such as fibrin, thrombin, albumin and gelatin, algal biopolymers pose no risk of infectious diseases transmission, (e.g. AIDS, Mad Cow Disease). Algae and algal products have long history of proven safety as food, medical and cosmetic products. Biodegradation of the algal polymers leads to non-toxic nutrient products. Large volume of cultivated and naturally sourced algal biomass is available. Mechanical and physico-chemical properties of algal biopolymers have been successfully modulated, e.g. by plasticization, cross-linking and combination with other natural and synthetic polymers.[77]

The most attractive biomedical applications of algal biopolymers include areas as broad as closure of surgical incisions, punctures, trauma-induced lacerations and skin graft attachment, surgical strips, wound dressings, scaffolds for tissue engineering, hydrogels, and carriers for drug delivery systems.[78] Apart from its primary function to bind and join tissues, the algal adhesive can also perform a variety of secondary functions such as to stop wound bleeding and facilitate healing processes.

Attachment of living tissues is a very attractive yet complicated issue. An ideal tissue adhesive should comply with a number of special requirements. It should be viscous enough to be easily applied without excessive spreading on the tissue surface, it should adhere strongly to moist or even wet tissues at body temperature and be flexible to conform to the morphology of the adhered tissues. Moreover, it should be biocompatible and biodegradable *in vivo*, possess neither allergenic nor carcinogenic nor mutagenic properties and be stable to ensure adequate shelf life. Also, the by-products formed during adhesive biodegradation should be eliminated from the organism by common means, without accumulation in organs or tissues.[79,80]

Successful bioadhesive matrices for topical administration of active agents for prolonged periods of time (controlled release systems) should maintain intimate contact with the site of application for the whole period of administration, be sufficiently adhesive and cohesive, guarantee controlled delivery of the active ingredients in wet and moist environments, be non-toxic, non irritating and easily removable. Many active agents can be released via bioadhesive matrices, e.g. steroids, anti-inflammatory agents, pain relief drugs, pH-sensitive peptides and small proteins such as insulin.[81]

The most common algal-derived bioadhesives are based on algal polysaccharides such as alginate, agar and carrageenans.[19]

Alginate, an anionic polysaccharide present in cell walls of brown algae, is one of the most promising and useful polysaccharides with algal origin having a broad range of biomedical applications. Alginate hydrogels were employed as wound dressings promoting wound healing,[82] haemostatic agents for cavity wounds, absorbent for heavily exudating wounds,[83] scaffold materials for tissue engineering[84,85] and controlled-release vehicles for pharmaceutical applications.[19]

In tissue engineering alginates has a great potential.[86-88] Alginate based scaffolds have been shown to be compatible with and support the growth of various types of cells as diverse as chondrocytes,[89] fibroblasts,[90] osteoblasts,[67,91] hepatocytes[92,93] and pancreatic islets.[94,95] Alginate scaffolds can be produced in desired shapes and structures, e.g. as macroporous beads with an highly interconnected pore structure,[78] as a composite material made of separate deposited layers,[85] in the shape of facial implants[89] and for tympanic membrane patches.[96] The mechanical properties of the scaffold may be regulated by the type, molecular weight and concentration of the used alginate as well as by the chemistry and concentration of the alginate used, as well as the chemistry and concentration of the cross-linker used.[97-100]

The adhesive properties of the alginate substrate may be also tuned, e.g. by the controlled application of sodium citrate, a chelating agent typically used to retrieve cells from alginate cultures.[85] Mechanical properties and biodegradation behaviour of cross-linked alginate gels may be also tailored, for example by incorporation of hydrolytically degradable ester linkages and changing the biopolymer's cross-linking density.[101]

Biomimicry of the natural adhesive seems a promising route to replace natural adhesive materials and could be employed to formulate synthetic biomimetic adhesives. In fact, production of natural algal bioadhesives relies on the complex and time consuming process of harvesting algal plants followed by extraction and purification of the active ingredients. A biomimetic adhesive based on alginates, calcium and phloroglucinol was recently proposed by Bitton and co-workers.[102,103] The addition of phloroglucinol and calcium ions to the alginate allows the formation of a three dimensional network able to promote tissue adhesion in aqueous environments.

Mucoadhesive films for buccal drug delivery based on sodium alginate capsules were described by Skulason *et al.*[104] Alginate films modified by hydroxypropylmethyl cellulose and propylene glycol as plasticizer possessed bioadhesive properties as well as tensile strength and elasticity higher than those of films made of the synthetic bioadhesive polymer Carbopol. The release profile of a model drug, sumatriptan succinate, showed that the drug release mechanism was diffusion controlled rather than associated to disintegration of the films. The results indicated that sodium alginate may be a suitable carrier for use in the buccal cavity.

Agar is another algal polysaccharide having biomedical and pharmaceutical applications with industrial relevance. Agar is used as suspending agent for radiological solutions, as a formative ingredient for tablets and capsules for drug delivery and release[19] and as skin dressing for wounds.[105] Agar is also the medium of choice for culturing bacteria on solid substrates. Algal polysaccharides carrageenans were also found attractive for production of wound dressings in combination with synthetic polymers[106] and as moisturizers and active carriers of drugs.[19]

5   CONCLUSIONS

The composition and adhesion process of algal bioadhesives are largely unknown and have yet to be elucidated. Preliminary studies of bioadhesives and biomimetic analogues demonstrated the feasibility of using algal products in biomedical applications, including surgical tissue adhesives, cell attachment enhancers and skin and buccal delivery of active agents.

In this chapter, the biomechanical, chemical and biocompatibility properties of a novel bio-adhesive extract from the red-algae *Gracilaria conferta* were discussed. Plasticization and cross-linking were considered to be useful post-treatments to fine tune the characteristics of the algal product, allowing the manufacture of strong and brittle as well as elastic bio-materials from the same algal source. The mechanical properties of *G. conferta* films were comparable to those of commercial films for wound and burn dressing. In addition, *G. conferta* extracts were able to effectively bind hydrophobic materials with adhesion strength comparable to a cyanoacrylate based glue. Biocompatibility of the bio-adhesive was confirmed by a number of *in vivo* and *in vitro* tests. Thanks to the easy fine tuning of the physical properties of the algal extract coupled to its good mechanical and biocompatibility characteristics, the bioadhesive presented in this work shows good promise to be used in the biomedical arena.

As for all other bio-based adhesives, algal bioadhesives are complex and extensive multidisciplinary research is still required to elucidate the bioadhesion mechanism, with the ultimate aim of bringing innovative bio-inspired adhesives to the market.[107]

References

1   J. Dove and P. Sheridan, *J. Am. Dent. Assoc.*, 1986, **112**, 879.
2   C. Saez, J. Pardo, E. Gutierrez, M. Brito and L.O. Burzio, *Comp. Biochem. Phys. B*, 1991, **98**, 569.
3   L. Ninan, R.L. Stroshine, J.J. Wilker and R. Shi, *Acta Biomater.*, 2007, **3**, 687.
4   M. Wiegemann, T. Kowalik and A. Hartwig, *Mar. Biol.*, 2006, **149**, 241.
5   A. Majumder, A. Sharma and A. Ghatak, *Langmuir*, 2009, **26**, 521.
6   H.J. Cha, D.S. Hwang and S. Lim, *Biotech. J.*, 2008, **3**, 631.
7   H. Lee, S.M. Dellatore, W.M. Miller and P.B. Messersmith, *Science*, 2007, **318**, 426.
8   L. Khandeparker and A.C. Anil, *Int. J. Adhes. Adhes.*, 2007, **27**, 165.
9   H. Shao, K.N. Bachus and R.J. Stewart, *Macromol. Biosci.*, 2009, **9**, 464.

10 H. Lee, N.F. Scherer and P.B. Messersmith, *P. Natl. Acad. Sci. USA*, 2006, **103**, 12999.

11 C.L. Stevens, C.L. Hurd and M.J. Smith, *J. Exp. Mar. Biol. Ecol.*, 2002, **269**, 147.

12 D.L. Harder, C.L. Hurd and T. Speck, *Am. J. Bot.*, 2006, **93**, 1426.

13 D.I. Taylor and D.R. Schiel, *J. Exp. Mar. Biol. Ecol.*, 2003, **290**, 229.

14 R. Bitton, M. Ben-Yehuda, M. Davidovich, Y. Balazs, P. Potin, L. Delage, C. Colin and H. Bianco-Peled, *Macromol. Biosci.*, 2006, **6**, 737.

15 J.A. Callow, S.A. Crawford, M.J. Higgins, P. Mulvaney and R. Wetherbee, *Planta*, 2000, **211**, 641.

16 V. Vreeland, J.H. Waite and L. Epstein, *J. Phycol.*, 1998, **34**, 1.

17 N. Fusetani, *Nat. Prod. Rep.*, 2004, **21**, 94.

18 M.E. Callow and J.A. Callow, *Biologist*, 2002, **49**, 1.

19 P. Laurienzo, *Mar. Drugs*, 2010, **8**, 2435.

20 V.K. Dhargalkar and X.N. Verlecar, *Aquaculture*, 2009, **287**, 229.

21 A.S. Carlsson, J.B. van Beilen, R. Moller and D. Clayton, *Micro- and macro-algae: Utility for industrial applications*, CPL Press, Newbury, 2007.

22 M. Peyriere, *C. R. Acad. Sci.*, 1970, **270**, 2071.

23 D.L. McBride and K. Cole, *Phycologia*, 1971, **10**, 49.

24 J.R.J. Baker and L.V. Evans, *Protoplasma*, 1973, **77**, 1.

25 M.E. Callow and L.V. Evans, *Protoplasma*, 1974, **80**, 15.

26 L. Oliveira, D.C. Walker and T. Bisalputra, *Protoplasma*, 1980, **104**, 1.

27 M.S. Stanley, M.E. Callow and J.A. Callow, *Planta*, 1999, **210**, 61.

28 J.A. Callow, M.S. Stanley, R. Wetherbee and M.E. Callow, *Biofouling*, 2000, **16**, 141.

29 M.E.A. Schoenwaelder and C. Wiencke, *Plant Biology*, 2000, **2**, 24.

30 M.E. Pettitt, S.L. Henry, M.E. Callow, J.A. Callow and A.S. Clare, *Biofouling*, 2004, **20**, 299.

31 A.J. Humphrey, J.A. Finlay, M.E. Pettitt, M.S. Stanley and J. A. Callow, *J. Adhesion*, 2005, **81**, 791.

32 S. La Barre, P. Potin, C. Leblanc and L. Delage, *Mar. Drugs*, 2010, **8**, 988.

33 B. Levi and M. Friedlander, *J. Appl. Phycol.*, 2004, **16**, 1.

34 V.T. Wagner, L. Brian and R.S. Quatrano, *P. Natl. Acad. Sci. USA*, 1992, **89**, 3644.

35 W.R. Fagerberg, J. Towle, C.J. Dawes and A. Böttger, *J. Phycol.*, 2012, **48**, 264.

36 V. Vreeland and E. Grotkopp, *US Pat.*, 5,520,727, 1996.

37 P.M. Favi, S. Yi, S.C. Lenaghan, L. Xia, and M. Zhang, *J. Adhes. Sci. Technol.*, 2012, DOI: 10.1080/01694243.2012.691809.

38 E.L. McCandless and J.S. Craigie, *Ann. Rev. Plant Physio.*, 1979, **30**, 41.

39 G. Michel, W. Helbert, R. Kahn, O. Dideberg and B. Kloareg, *J. Mol. Biol.*, 2003, **334**, 421.

40 Z.L. Bouzon, L.C. Ouriques and E.C. Oliveira, *J. Appl. Phycol.*, 2006, **18**, 287.

41 Z.L. Bouzon and L.C. Ouriques, *Aquat. Bot.*, 2007, **86**, 301.

42 M.E. Apple and M.M. Harlin, *Phycologia*, 1995, **34**, 417.

43 M.E. Apple, M.M. Harlin, and J.H. Norris, *Phycologia*, 1996, **35**, 245.

44 L. Petrone, R. Easingwood, M.F. Barker and A.J. McQuillan, *J. R. Soc. Interface*, 2011, **8**, 410.

45 M. Haber, *US Pat.*, 5,859,198, 1999.

46 O.H. Lowry, N.J. Rosebrough, A.L. Farr and R.J. Randall, *J. Biol. Chem.*, 1951, **193**, 265.

47 K. Wilson and J. Walker, *Practical Biochemistry: Principles and Techniques*, Cambridge University Press, Cambridge, 5th edn., 2000.

48 S. Moore and W.H. Stein, *Methods in Enzymology*, Academic Press, 1963, 819.

49 S. Moore, D.H. Spackman and W.H. Stein, *Anal. Chem.*, 1958, **30**, 1185.

50 D. Beauchemin, *Anal. Chem.*, 2006, **78**, 4111.

51 B. Ray and M. Lahaye, *Carbohyd. Res.*, 1995, **274**, 251.

52 Z. Dische, *J. Biol. Chem.*, 1950, **183**, 489.

53 M.F. Chaplin and J.F. Kennedy, *Carbohydrate analysis: a practical approach*, IRL Press, Oxford, 1986.

54 T.T. Terho and K. Hartiala, *Anal. Biochem.*, 1971, **41**, 471.

55 H.F. Mark, N. Bikales, C.G. Overberger, G. Menges and J.I. Kroschwitz, *Encyclopedia of Polymer Science and Engineering, Volume 7: Fibers, Optical to Hydrogenation*, 2nd ed., Wiley-Interscience, New York, 1987.

56 *Standard Test Method for Tensile Properties of Thin Plastic Sheeting*, ASTM International, 2010.

57 *Standard Test Method for Tensile Properties of Adhesive Bonds*, ASTM International, 2008.

58 *Biological evaluation of medical devices - Part 5: Tests for in vitro cytotoxicity*, ISO 10993-5 standard, 2009.

59 *Biological evaluation of medical devices - Part 10: Tests for irritation and skin sensitization*, ISO 10993-10 standard, 2010.

60 *Biological evaluation of medical devices - Part 11: Tests for systemic toxicity*, ISO 10993-11 standard, 2006.

61 K.T. Preissner, *Annu. Rev. Cell Biol.*, 1991, **7**, 275.

62 E. Hennebert, R. Wattiez, J.H. Waite and P. Flammang, *Biofouling*, 2012, **28**, 289.

63 J.H. Waite and X. Qin, *Biochemistry*, 2001, **40**, 2887.

64 L. C. Ouriques, É. C. Schmidt and Z. L. Bouzon, *Micron*, 2012, **43**, 269.

65 A. Rosenhahn, J. Finlay, M. Pettit, A. Ward, W. Wirges, R. Gerhard, M. Callow, M. Grunze and J. Callow, *Biointerphases*, 2009, **4**, 7.

66 L. Petrone and A.J. McQuillan, *Appl. Spectrosc.*, 2011, **65**, 1162.

67 Z. Li, H.R. Ramay, K.D. Hauch, D. Xiao and M. Zhang, *Biomaterials*, 2005, **26**, 3919.

68 M.J. Sever and J.J. Wilker, *Dalton T.*, 2006, 813.

69 M. Avella, E.D. Pace, B. Immirzi, G. Impallomeni, M. Malinconico and G. Santagata, *Carbohyd. Polym.*, 2007, **69**, 503.

70 Y.S. Choi, S.R. Hong, Y.M. Lee, K.W. Song, M.H. Park and Y.S. Nam, *J. Biomed. Mater. Res.*, 1999, **48**, 631.

71 H.M. Powell and S.T. Boyce, *Biomaterials*, 2006, **27**, 5821.

72 V. Karageorgiou, L. Meinel, S. Hofmann, A. Malhotra, V. Volloch and D. Kaplan, *J. Biomed. Mater. Res. A*, 2004, **71A**, 528.

73 J.W. Rhim, A. Gennadios, A. Handa, C.L. Weller and M.A. Hanna, *J. Agr. Food Chem.*, 2000, **48**, 4937.

74 O. Ben-Zion and A. Nussinovitch, *Food Hydrocolloid.*, 1997, **11**, 429.

75 W.D. Spotnitz and S. Burks, *Transfusion*, 2012, **52**, 2243.

76 A.P. Duarte, J.F. Coelho, J.C. Bordado, M.T. Cidade and M.H. Gil, *Progr. in Pol. Sci.*, 2012, **37**, 1031.

77 G.G. d' Ayala, M. Malinconico and P. Laurienzo, *Molecules*, 2008, **13**, 2069.

78 P. Eiselt, J. Yeh, R.K. Latvala, L.D. Shea and D.J. Mooney, *Biomaterials,* 2000, **21**, 1921.

79 H.T. Peng and P.N. Shek, *Expert Rev. Med. Devic.*, 2010, **7**, 639.

80 T.E. Lipatova, *Adv. Pol. Sci.*, 1986, **79**, 65.

81 I. Webster and P.J. West in *Polymeric biomaterials*, ed. S. Dumitrion, 2nd ed., Marcel Dekker, New York, 2002, 703.

82 G.J. Motta, *Ostomy Wound Manag.*, 1989, **25**, 52.

83 T. Gilchrist and A.M. Martin, *Biomaterials*, 1983, **4**, 317.

84 K.Y. Lee and D.J. Mooney, *Chem. Rev.*, 2001, **101**, 1869.

85 J.P. Gleghorn, C.S.D. Lee, M. Cabodi, A.D. Stroock and L.J. Bonassar, *J. Biomed. Mat. Res. A*, 2008, **85A**, 611.

86 A. Dar, M. Shachar, J. Leor and S. Cohen, *Biotech. Bioeng.*, 2002, **80**, 305.

87 J. Bierwolf, M. Lutgehetmann, S. Deichmann, J. Erbes, T. Volz, M. Dandri, S. Cohen, B. Nashan and J.M. Pollok, *Tissue Eng. Pt. A*, 2012, **18**, 1443.

88 Y. Sapir, O. Kryukov and S. Cohen, *Biomaterials*, 2011, **32**, 1838.

89 S.C.N. Chang, J.A. Rowley, G. Tobias, N.G. Genes, A.K. Roy, D.J. Mooney, C.A. Vacanti and L.J. Bonassar, *J. Biomed. Mat. Res.*, 2001, **55**, 503.

90 S. Ponce, G. Orive, A.R. Gascón, R.M. Hernández and J.L. Pedraz, *Int. J. Pharm.*, 2005, **293**, 1.

91 D.J. Park, B.H. Choi, S.J. Zhu, J.Y. Huh, B.Y. Kim and S.H. Lee, *J Cranio Maxill. Surg.*, 2005, **33**, 50.

92 G. Mai, N.T. Huy, P. Morel, J. Mei, A. Andres, D. Bosco, R. Baertschiger, C. Toso, T. Berney, P. Majno, G. Mentha, D. Trono and L.H. Buhler, *Xenotransplantation*, 2005, **12**, 457.

93 J.H. Lee, D.H. Lee, J.H. Son, J.K. Park and S.K. Kim, *J. Microbiol. Biotechnol.*, 2005, **15**, 7.

94 Y.C. Song, Z.Z. Chen, N. Mukherjee, F.G. Lightfoot, M.J. Taylor, K.G. Brockbank and A. Sambanis, *Transpl. P.*, 2005, **37**, 253.

95 N.E. Simpson, N. Khokhlova, J.A. Oca-Cossio, S.S. McFarlane, C.P. Simpson and I. Constantinidis, *Biomaterials*, 2005, **26**, 4633.

96 M.E. Hott, C.A. Megerian, R. Beane and L.J. Bonassar, *Laringoscope*, 2004, **114**, 1290.

97 M.A. LeRoux, F. Guilak and L.A. Setton, *J. Biomed. Mat. Res.*, 1999, **47**, 46.

98 C.K. Kuo and P.X. Ma, *Biomaterials*, 2001, **22**, 511.

99 J.L. Drury, R.G. Dennis and D.J. Mooney, *Biomaterials*, 2004, **25**, 3187.

100 N.G. Genes, J.A. Rowley, D.J. Mooney and L.J. Bonassar, *Arch. Biochem. Biophys.*, 2004, **422**, 161.

101 O. Jeon, K.H. Bouhadir, J.M. Mansour and E. Alsberg, *Biomaterials*, 2009, **30**, 2724.

102 R. Bitton, E. Josef, I. Shimshelashvili, K. Shapira, D. Seliktar and H. Bianco-Peled, *Acta Biomater.*, 2009, **5**, 1582.

103 R. Bitton and H. Bianco-Peled, *Macromol. Biosci.*, 2008, **8**, 393.

104 S. Skulason, M.S. Asgeirsdottir, J.P. Magnusson and T. Kristmundsdottir, *Pharmazie*, 2009, **64**, 197.

105 L. Bao, W. Yang, X. Mao, S. Mou and S. Tang, *Biomed. Mater.*, 2008, **3**, 1.

106 A.B. Lugão, L.D.B. Machado, L.F. Miranda, M.R. Alvarez and J.M. Rosiak, *Radiat. Phys. Chem.*, 1998, **52**, 319.

107 M. Gross, *Chemistry World*, 2011, **8**, 52.

BACTERIAL ADHESION SYSTEMS AS AN ALTERNATIVE FOR THE INDUSTRY: A CASE STUDY

N. Cuesta-Garrote, M.J. Escoto-Palacios*, F. Arán-Ais and C. Orgilés-Barceló

Footwear Technological Institute-INESCOP, Polígono Industrial Campo Alto, Elda (Alicante), Spain.
*mjescoto@inescop.es

1 BACKGROUND: INDUSTRIAL ADHESIVES CONSTRAINTS

Until the twentieth century, natural materials were the predominant basis for most adhesives. However, the emergence of synthetic polymers in the early 1930s led to a significant change in adhesive formulations. In fact, in the period before the Second World War, new polymers were crucial for the modern adhesives industry,[1] whose products turned to be based on non-biodegradable synthetic polymers from fossil sources, and to use organic solvents as carriers.

This change in adhesive formulations was reflected in industrial manufacturing processes, where the adhesives were introduced in the assembly operations. It allowed headway to be made towards a more standardised production process, with higher quality finished products, therefore acquiring a less traditional and more industrialised nature.[1,2] However, despite the technical benefits brought by synthetic adhesives, the use of organic solvent and fossil raw material based adhesives implies a number of risks, among which their environmental impact and adverse effects on human health should be highlighted.[3,4]

European legislation related to environmental protection,[5,6] industrial safety and the health of workers[7] has led to the reduction and/or avoidance of many organic solvents used in the footwear industry as well as the reduction of non-biodegradable waste. Specifically, Directive 2010/75/EU of the European Parliament and the Council on industrial emissions, establishes the emission limit values for volatile organic compounds due to industrial activities. Moreover, Directive 2008/98/EC on waste obliges Member States to take the necessary measures to reduce waste production through the development of clean technologies and encouraging waste recycling.

These increasing legislative requirements, along with a growing consumer environmental awareness, makes industries require new production systems that are friendlier to the environment, without having a loss of competitiveness. As a result, solvent-free alternative technologies, such as hot-melt formulations and waterborne adhesives have been developed[8] and are already being used by some industries. However, they do not provide a solution to the environmental burden derived from the non-biodegradability of adhesive components. For this reason it is necessary to develop sustainable and environmentally-friendly industrial adhesives.

In fact, there is a growing interest in biobased materials at both a governmental,[9,10] and industrial level. According to a market study of the industry association European

Bioplastics, the production of bioplastics in 2010 was around 700,000 tonnes, and is expected to reach 1.7 million tonnes in 2015.[11] However, very little information has been found about biomaterial applications in adhesive formulations for the industry.

2 SEEKING ALTERNATIVES: ADHESION MECHANISMS IN NATURE

Different bonding systems have been observed in nature that allow living organisms to adhere firmly to different surfaces, or to capture their nutrients, even in highly adverse conditions where other conventional synthetic adhesives would present problems. This is the case of gecko, some arachnids, crustaceans or molluscs, as well as microorganisms like those studied by the authors of this review.

Such adhesion mechanisms can be of quite different nature: Van der Waals interactions occur in the attachment system of gecko, whereas natural glues participate in adhesion mechanisms in molluscs and crustaceans, spider webs, and microorganisms.

2.1 Adhesion without Adhesives

In some organisms, adhesion occurs without the participation of adhesive substances, but due to the complex hierarchical structure of their feet.

This is the case of geckos, whose attachment capability lies in the structure and function of their feet (see Figure 1). Each digit has a toe pad consisting of a series of modified lamellae, each one covered with uniform arrays of similarly oriented hair-like β-keratin bristles called setae. Each seta splits into hundreds of 200 nm wide spatular tips. This hierarchical structure leads to a maximized number of contact points with the substrate, giving rise to high van der Waals dispersion interactive forces per unit area.[12-15]

A similar attachment system can also be found in other organisms, such as insects and arachnids.[16,17] In these organisms, however, the attachment pads are supplemented with various kinds of fluids. Attractive capillary forces mediated by these fluids are an important factor in the mechanism of insect attachment in addition to van der Waals forces.[18]

Gecko-like attachment systems have been deeply studied by several researchers with the aim of mimicking their extraordinary climbing abilities. The distribution and pattern of the setae arrays differ depending on the species, and some studies have shown that there is an inverse scale effect: the heavier the animal, the smaller and more densely packed the tips.[16,17]

2.2 Natural Glues

Naturally produced adhesives are common in many biological systems and are known for their superior strength and durability compared with man-made materials. Examples of specialised biological systems that generate a vast amount of research on their adhesives include bacteria, spider webs, marine tubeworms, sea cucumbers, barnacles, mussels, and plants.

2.2.1 Adhesives not for Self-attaching Purposes. Natural adhesives are not only produced by organisms for the purpose of attaching themselves to a surface. Different biopolymers can also be secreted by predator organisms with the aim of trapping their preys, or by preys as a defence strategy.

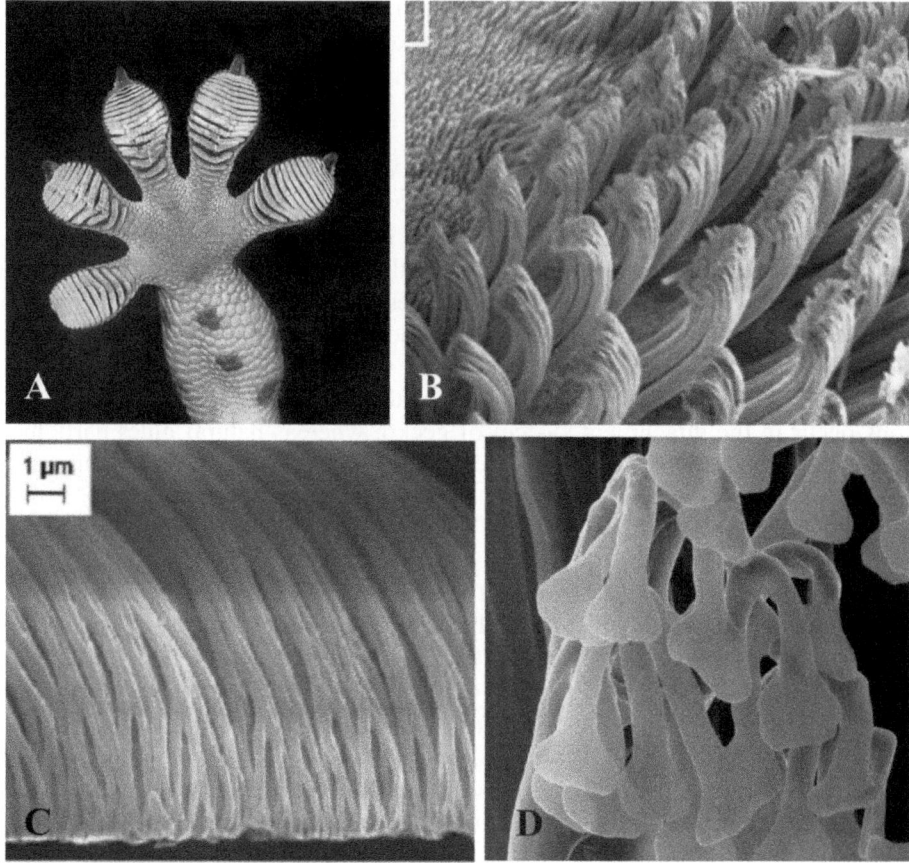

Figure 1   *Hierarchical attachment system in a gecko. A: Gecko's foot (Credit: David Clements); B: Array of setae in gecko's toe pads (Credit: Laboratory of Bio-inspired Nanomechanics "Giuseppe Maria Pugno", Politecnico di Torino); C: Spatular tips in a single seta. (Credit: Laboratory of Bio-inspired Nanomechanics "Giuseppe Maria Pugno", Politecnico di Torino); D: Spatula pads (Credit: Carl Zeiss).*

Leafs in some carnivorous plants have a series of glands producing sticky secretions whose composition varies with the species. *Drosera, Drosophyllum, Pinguicula* and their relatives produce a glue based on a polysaccharide mucilage, whereas the adhesive secreted by *Roridula* is based on terpenoid resins.[19]

Spider silks that make up the web of the orb-web spiders have been deeply studied due to their unique mechanical properties. They indeed present high tensile strength, extensibility and an energy-dissipative viscoelastic response that is hardly matched by synthetic polymers.[20,21] However, the effectiveness of spider webs in trapping and retaining preys also lays on an aqueous glue that coats the sticky-spiral threads of the web. This glue, which is considered to be one of the strongest and most effective biological glues, is an aqueous solution of glycoproteins. A study on the aqueous glue secreted by the golden orb weaving spider *Nephila clavipes* has shown that the glue glycoprotein is formed by two protein subunits that are encoded by opposite strands of the same DNA sequence and contain unique 110 amino acid repetitive domains.[22]

Figure 2 *Use of adhesives with defensive purposes. A: Tubeworm* P. californica *making a tube in the laboratory (Credit: Fred Hayes for the University of Utah); B: Sea cucumber* H. forskali *expelling Cuvierian tubules (Credit: Roberto Pillon).*

Conversely, other organisms use adhesives as a part of their defensive strategies (see Figure 2). The marine polychaete *Phragmatopoma californica* is a tubeworm that builds protective "tubes" with secreted proteinaceous cement mixed with shell fragments and sand particles from the sea floor. The cement adheres rapidly to a variety of materials in seawater.[23,24] Sea cucumbers (*Holothuria forskali*) react defensively through the ejection of sticky Cuvierian tubules that ensnare their predators. The biochemical analysis of the tubule's glue indicates a 3:2 protein to carbohydrate ratio, with a high proportion of highly insoluble protein. The soluble protein component appears to contain up to ten glycine- and acidic amino acid-rich proteins ranging from 17 to 220 kDa in size.[25] Furthermore, adhesive proteins in both species show some similarities in their amino acid composition, such as the presence of post-translationally modified amino acid phosphoserine (pSer), which has been reported to provide both adhesive and cohesive properties.[26] In addition, the tubeworm cement contains another modified amino acid, 3,4-dihydroxyphenylalanine (DOPA),[23,24,27] which appears infrequently in proteins but is also present in other protein-based adhesives. DOPA has been reported to participate in the crosslinking of the tubeworm cement [23,27] and, therefore, improves its cohesive strength. Both production of DOPA and cement crosslinking may be catalyzed by a co-secreted tyrosinase.[27]

2.2.2 High Performance Attachment: Molluscs and other Marine Organisms. Many marine organisms have developed adhesive strategies to deal with the dynamic ocean environment, particularly in the intertidal area.

Aquatic bacteria *Caulobacter crescentus* (Figure 3A) adhere to surfaces by using an extremely strong, polar adhesin called the "holdfast", which is located at the tip of its polar stalk, a thin cylindrical extension of the cell membrane. The holdfast of *C. crescentus* is composed of extracellular polysaccharides and other components such as proteins, and is considered to show the strongest adhesive strength observed in bacteria.[28-31] More recently, studies have showed that this holdfast is a complex of polysaccharides anchored to the cell by the surface-exposed outer proteins HfaA, HfaB and HfaD.[32] Two major genetic loci have been identified that are directly associated with the function and biosynthesis of holdfast polysaccharide, the *hfs* (holdfast synthesis) and *hfa* (holdfast anchoring) loci. The *hfs* locus encodes proteins involved in the biosynthesis (cluster *hfsEFGH*) and secretion (cluster *hfsDABC*) of the holdfast polysaccharide.[31,32] The *hfa* locus (operon *hfaABD*) is important for attaching the holdfast polysaccharide to *C. crescentus* cells.

Molluscs, such as the blue mussel *Mytilus edulis*, attach themselves to highly irregular surfaces by means of a bundle of fibrous threads ending in adhesive plaques or holdfast, the so-called byssus (Figure 3B). The adhesive plaque is mainly composed of collagen, catechol oxidase, as well as a series of adhesive proteins of polyphenolic nature (called Foot Proteins) that are considered to be crucial in adhesion mechanisms.[33-39] Foot Proteins are characterised by the repetition of oligopeptide motifs, by the abundance of lysine residues in their primary structure (about 20%) and because they undergo a series of post-translational modifications, leading to mono- and di-hydroxylated aminoacids (about 50%) consisting of hydroxyarginine, hydroxyproline, dihydroxyproline and DOPA. Furthermore, good homology has been found between adhesive foot proteins among different *Mytilus* species.[40] As in the case of the tubeworm cement, the presence of polyphenolic coupling products such as 5,5'-diDOPA in mussel byssal plaques[41] is considered to mean that DOPA, in its oxidized dopaquinone form, is responsible for crosslinking in these adhesive proteins, acting as an internal "hardener", which enables them to form highly resistant bonds even under highly severe conditions.[42,43] When DOPA is in its reduced form, hydroxyl groups can also form strong interactions with polar surfaces, therefore participating in adhesion mechanisms. However, such interactions are highly dependent on a redox modulation and compete with cross-linking process.[44,45]

Barnacles (Figure 3C), the only sessile crustaceans, adhere directly and permanently to various surfaces, such as rocks and several man-made substrata, including ship hulls, oil platforms, and pipelines. Initial attachment of barnacle cyprid larvae is via o-quinone cross-linking that resembles the DOPA-containing adhesive proteins of *Mytilus spp*.[46] However, the adult barnacle cement is substantially different and is comprised of three groups of proteins that contain high levels of the amino acids serine, threonine, glycine, and alanine.[46,47] A pattern of short, alternating regions of hydrophobic and hydrophilic residues throughout the largest of the three groups of proteins has been noted, suggesting that these alternating motifs may have a role in assembly in seawater.[47]

Figure 3   *Attachment systems in marine organisms. A:* Caulobacter crescentus *(Credit: Yves Brun, Indiana University); B: Mussel byssus (Credit: Brocken Inaglory, via Wikimedia Commons); C: Barnacles attached to a rock. (Credit: Auguste Le Roux).*

2.2.3 Bacterial Biofilms. It is known that non-aquatic microorganisms are also capable of forming biofilms on different biotic and/or abiotic surfaces by means of different extracytoplasmic or extracellular biopolymers, including proteins.

This is the case of *Pseudomonas* [48-51] (Figure 4A), which are organisms that can be found either in soils[50] or in clinical isolates.[51] More specifically, in the case of the non pathogenic organism *Pseudomonas fluorescens* WCS365, it is known that they contain the so-called *lap* genes, which encode an ABC transporter whose final secretion product is Lap A protein. The Lap A protein is a surface protein acting as a multifunctional adhesin that is organised in four domains. Two of them (domains 2 and 3), as occurs in mussel foot proteins, are composed of repeated motifs. Domain 2 consists of nine quasi-perfect repeats of 100 amino acids. Domain 3 is a large repetitive region, spanning 6400 amino acids, organized in 29 imperfect repeats of 218-225 amino acids. Good homology has been found for this protein among different *Pseudomonas* species.[50]

Other organisms, such as oral bacteria *Streptococcus* (Figure 4B), show adhesion mechanisms in which a series of adhesins called fimbriae are implied. Fimbriae mainly consist of structural proteins and lipoproteins present in the extracellular medium. The receptors of these fimbriae or adhesins are, in most cases, polysaccharides associated with the cell wall.[52] The literature describes 5 different types of fimbriae, which are responsible for the selective bonding of different streptococcus species to certain substrates, mainly dental plaque in a co-aggregation mechanism where fimbriae recognise complementary carbohydrate molecules on the partner's cell.[53-59]

Actinomyces (Figure 4C) are another type of Gram-positive bacteria involved in biofilm formation in dental plaque by means of mechanisms of co-aggregation with other bacteria. They attach to the surfaces to be colonised using two different types of fimbriae that recognise different signals. Type 1 fimbriae interact with salivary proteins that coat the enamel surfaces, whereas type 2 fimbriae participate in adherence to the cell wall polysaccharide of oral streptococci or to the oligosaccharides on mammalian cells.[60-64]

According to the data obtained using the BLASTP tool for the alignment of amino acid sequences, our research group found that for both actinomyces fimbriae there is a certain homology with a *Bacillus cereus* ATCC14579 protein of similar molecular weight and involved in collagen adhesion.[65] *B. cereus* is a Gram-positive bacterium, with some of its strains being pathogens. Several strains have indeed a special surface structure called S-layer, consisting of a crystalline glycoprotein. This S-layer has a significant role in the pathogenic nature of this microorganism as it is involved in the adhesion to host cells. However, this surface structure has not been observed in *Bacillus cereus* ATCC14579 strains,[66,67] which have different collagen-binding adhesive proteins. Most of these proteins contain a leader peptide and thus they are expected to be extracellular proteins. Besides, some of them show domain repeats, as occurred in adhesive proteins from mussels and in the LapA protein from *Pseudomonas*.

The BLASTP tool was used within the GenoList integrated environment,[68-70] which includes an updated database of the whole genome of *Bacillus subtilis*. As a result, some proteins from this organism have been found to have homologies with the bacterial adhesive proteins previously described in this section.[65] *B. subtilis* (Figure 5D) is a well known natural inhabitant of the soil [71-75] whose genome has been completely sequenced.[70,76-78] *B. subtilis* forms rough biofilms at the air-liquid interface and is capable of producing high amounts of proteins, being considered as a potential industrial microorganism for the production of high quality proteins and enzymes.[79,80]

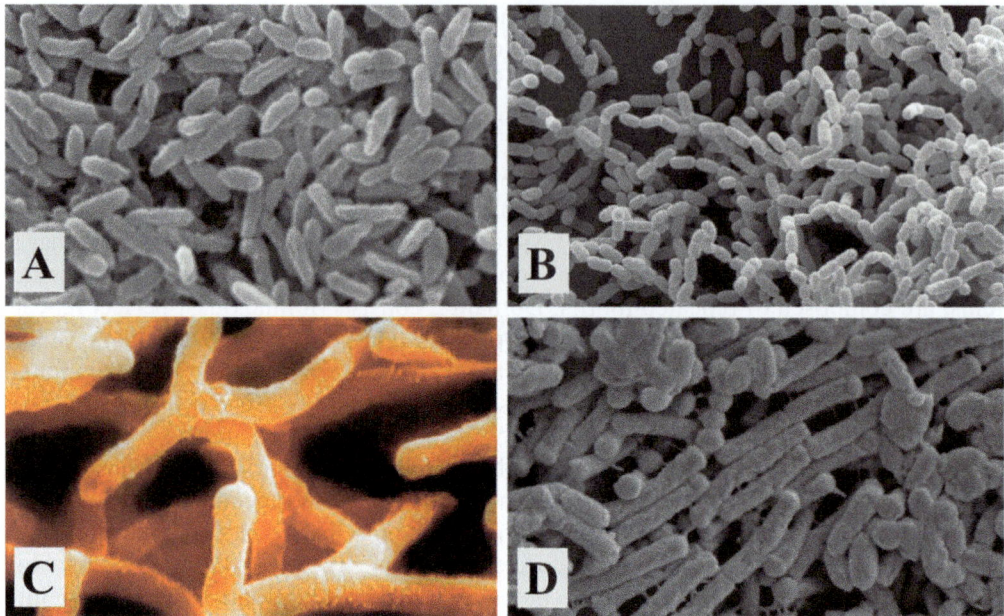

Figure 4   *Bacterial biofilms. A:* Pseudomonas aeruginosa. *(Credit: Deligianni et al.[51])*;
B: Streptococcus mutans *(Credit: UiO. Institute for oral biology); C:*
Actinomyces israelii *(Credit: Graham Colm); D:* Bacillus subtilis *(Credit:*
INESCOP).

3 MIMICKING NATURE: BIOINSPIRED PROPOSALS AS ALTERNATIVE ADHESIVES

In the last years, several researchers have focused their works in understanding, producing and/or mimicking the different attachment mechanisms found in nature, with the aim of developing new adhesion systems to be used in several areas such as medicine and industry.

Among all the natural attachment systems, those that have aroused the most interest among researchers have been, undoubtedly, those used by geckos and mussels.

Several approaches have been made in order to mimic the gecko's attaching mechanisms, which have been possible thanks to the advances in micro- and nanofabrication technologies.[81]

Prof. Messersmith's Research Group at Northwestern University in Illinois (USA) has developed some materials with adhesive properties that jointly mimic both mussel and gecko adhesion systems.[82] These materials have been proposed for their application as pressure-sensitive, removable adhesives in the medical field. However, the temporary nature of the bonds and their presentation in sheet form could restrict their use as industrial adhesives.

With regards to protein-based glues inspired from mussels and other marine organisms, the expected compatibility of protein adhesives with living tissues, the high resistance that some of them show to biodegradation and their capacity for bonding, even in an aqueous medium, have contributed to considering the study of their medical and surgical applications.[42,83-86] Professor H. Waite's Group, at the Department of Molecular, Cellular and Developmental Biology of the University of California at Santa Barbara,[33,37-

[39,41,43-45,86,87] the "Interface Biophysics" Group in the Cellular and Molecular Biology Department of Göteborg University [88,89] and Wilker's Group in the Chemistry Department of Purdue University in Indiana [83,90] are examples of researchers that have studied in depth mussel adhesive proteins and have worked on adhesives inspired by them. Furthermore, the Group of H. Silverman and F. Roberto in the Department of Biological Systems of the University of Idaho has developed a protocol for cloning, expressing and purifying *Mytilus edulis* foot proteins Mefp-1 and Mefp-2.[91,92]

These and other works led to some products being put onto the market,[93] such as "Mussel Adhesive Protein" from Sigma-Aldrich and "Cell-Tak" from BioPolymers Corporation,[94] which are based on adhesive protein extracts from *Mytilus edulis*. In addition, the Genex Corporation developed the product called "AdheraCell" [95] consisting of a recombinant protein obtained from the synthetic gene of Mefp-1. In all cases, the application of these products is restricted to laboratory use as systems for the fixation and adhesion of cells or tissue sections. Currently, the only commercially available product is "Cell-Tak", which is marketed by BD Biosciences, whereas the other products have been discontinued.

4 BACTERIAL ADHESIVES AS AN ALTERNATIVE FOR THE FOOTWEAR INDUSTRY

4.1 Alternative Adhesive Choice

In industrial manufacturing, adhesives play an important role during the assembly operations and are a key parameter to be taken into account when designing the item to be produced. An inadequate adhesive for a specific material will be a source of failures in the quality of the product and, therefore, of complaints.

In the particular case of footwear manufacturing, there are a large number of processes where several components made of materials from different origins, both natural and synthetic, must be joined together (see Figure 5). The high diversity of materials used in the footwear industry, as well as their steady evolution, often gives rise to adhesive bonding problems. For this reason, this development in materials has meant a parallel development in adhesives in the footwear sector.

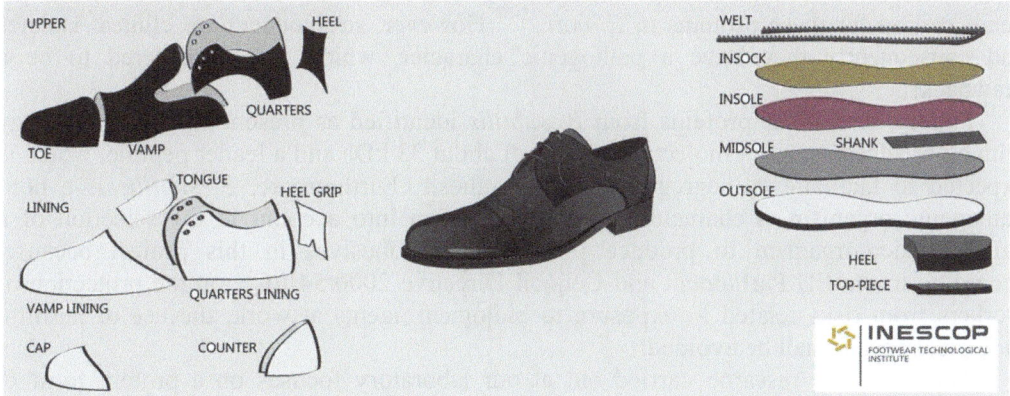

Figure 5 *Shoe components*

One of the operations that has the biggest influence on final product quality is the bonding of the upper to the sole. This operation is carried out using a wide range of adhesives, which are in each case adapted to the properties of the materials which are to be bonded, fulfilling the corresponding technical requirements of the upper-sole bond. Up to now, upper-to-sole bonding has been performed using mainly solvent-based contact adhesives. However, as mentioned before, these adhesives imply a series of environmental repercussions and can have destructive effects on human health.

In response to this need, in our laboratory we have worked on a research line focused on the adaptation of natural adhesives as those described above to be used as the base polymer for the formulation of industrial adhesives. Furthermore, microbial biotechnology and industrial microbiology procedures have been proposed to produce such biopolymers.

Even though proteins from marine organisms have proved to be of great interest for use as a polymer base for industrial adhesives, their production for commercial purposes were found to pose some difficulties. Some studies reported the chemical or biotechnological synthesis of peptides. However, at the moment we had to make our choice, all the studies referred to small peptides and not to whole adhesive proteins. Furthermore, there was no evidence of any test which would prove the adhesive capability of these peptides.[34,96] Additionally, microorganisms do not perform post-translational modifications that are needed to obtain DOPA, one of the amino acids supposed to be responsible for adhesive properties. These post-translational modifications should be subsequently produced *in vitro* by means of bacterial[42] and mushroom[87] tyrosinases. In addition, the high molecular weight of these proteins was considered to make the whole process difficult. On the other hand, the complex metabolic pathway to produce holdfast adhesive by *C. crescentus*, makes difficult its synthesis by a molecular biology approach.

In view of these limitations for the production of adhesives inspired by marine holdfasts, it was decided to work on bacterial proteins with known or potential adhesive properties. Both homologous and heterologous expression systems were considered as the way to produce them.

Adhesive proteins in *Pseudomonas* have the advantage that they do not undergo any post-translational modification. However, they are very large proteins, with a molecular weight of about 900 kDa, which could imply difficulties during their segregation once synthesized.

In the case of streptococci proteins, all the described adhesive lipoproteins have a molecular weight of around 35 kDa, therefore their segregation is expected to be easier once synthesized. There are also some studies referring to the cloning processes of the genes coding for these proteins in *E. coli*.[55-57] However, streptococci are clinical isolates and consequently they have a pathogenic character, which was considered to be a drawback.

Finally, one of the proteins from *B. subtilis* identified as presenting good homology with other adhesins, has a molecular weight of about 33 kDa and a leader peptide, which is expected to facilitate its segregation after synthesis. Furthermore, *B. subtilis* is a non-pathogenic organism, a characteristic that was taken into account in the selection of a suitable microorganism to produce protein-based adhesives in this project because, according to the EU Parliament and Council Directive 2000/54/EC[97] on the protection of workers from risks related to exposure to biological agents at work, the use of harmful biological agents shall be avoided.

Therefore, the research carried out at our laboratory focuses on a protein from *B. subtilis* that, according to a sequence alignment analysis (BLAST), shows a high percentage of homology with sequences of proteins with a known adhesive function.

4.2 Participation of the Selected Protein in Adhesion and Biofilm Formation

In a previous study, we have confirmed the involvement of the selected protein in biofilm formation mechanisms in *B subtilis*.[98]

In that work, the protein was highly produced in a *B. subtilis* overexpressing strain. Biofilm formation by both wild type and overexpressing strains was monitored using a modified version of the microtitre plate assay[98] that had been previously described by O'Toole and Kolter for *Pseudomonas fluorescens*[49] and adapted by Hamon and Lazazzera[99] for *B. subtilis*. Differences among both strains' behaviour were appreciable (see Figure 6A). Wild type strain preferably produced a biofilm at the air-liquid interface, whereas the biofilm produced by the overexpressing strain attached preferentially to the well surface, presenting an apparent longer-lasting affinity for the surface of the well over time.

The analysis of biofilm morphology in scanning electron microscopy showed that cells in biofilms appeared to be completely encased in an extracellular matrix when the protein was overproduced (Figure 6B). In addition, Confocal Laser Scanning Microscopy (CLSM) studies showed that while the adhered cells of the overexpressing strain were distributed as a three-dimensional, multicellular structure typical of a biofilm, in the wild type strain cells were dispersed along the surface (Figure 6C).

These results confirmed the relationship between the expression of the selected protein and an increase in biofilm production and adherence to abiotic surfaces and allow considering the use of this protein as a base polymer in the formulation of bioinspired adhesives.

4.3 Protein Characterisation and Suitability Assessment

In a recent study, the strain *Escherichia coli* M15 has been used as a heterologous expression system to express the protein of interest. The adhesive protein under study has been subcloned into a commercial expression vector (pQE60) inducible by IPTG that incorporates a six-histidine tag to the protein at its carboxyl terminus. This tag favours subsequent purification of the protein by affinity chromatography by using a matrix system of nickel-nitrilotriacetic acid resin (Ni - NTA).

Once the protein has been purified, its suitability for use as a polymeric base in adhesive formulations has been assessed. Properties studied include average particle size,

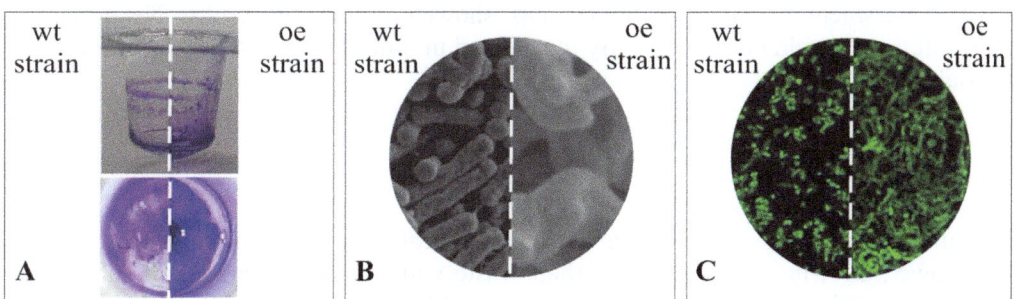

Figure 6 *Overview of the involvement of the protein under study in biofilm formation by* B. subtilis. *Comparison of the behaviour of the wild type strain (wt strain) vs the overexpressing strain (oe strain). A: Microtitre plate assay, B: Biofilm morphology analysis by SEM, C: Biofilm morphology analysis by CLSM.*

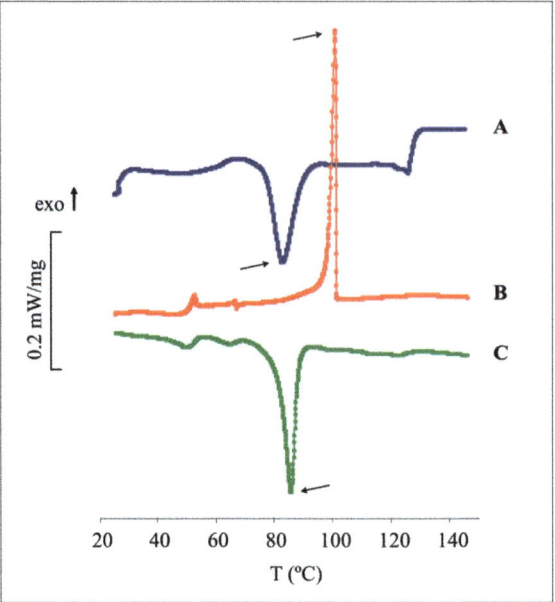

Figure 7   *DSC thermograms of the purified protein. Heating and cooling scans obtained under a N_2 stream. A: First heating scan, B: cooling scan, C: second heating scan. Arrows point at a reversible denaturing process.*

wetting behaviour, surface energy, thermal properties and thermal stability, among others.[100]

An interesting finding has been provided by the analysis of the thermal properties of the lyophilized purified protein. The DSC thermograms obtained under a N_2 stream showed a reversible denaturing process at about 70°C (see Figure 7). This temperature is close to the reactivation temperature of conventional polyurethane adhesives used in footwear manufacturing. As a result, it has been concluded that the selected protein could be explored as a heat-reactivated adhesive.

Furthermore, contact angle measurements of an aqueous solution of the protein on conventional soling materials have shown contact angle values under 45°, indicating a good wetting behaviour on such substrates.

The assessment of other properties has showed promising results. Nevertheless, protein solutions should be conveniently formulated in order to improve its performance as industrial adhesives.

5 CONCLUSIONS

The nature of protein-based adhesives seems to indicate that their environmental impact will be lower, allowing the disposal of their waste without the appearance of substances that may be considered harmful to the environment. Moreover, the amino acid composition and the side chains of the folding structure allow these proteins to be easily solubilised in aqueous media. In addition, their unique characteristics of adhesion and resistance to harsh environments make them suitable to explore their possible applications as adhesives in the industry.

These results open up the possibility of considering the use of bacterial adhesive proteins as the polymeric base for the formulation of industrial adhesives as a sustainable and competitive alternative to conventional adhesives used in manufacturing, leading to the production of distinctive items with an added value.

From this point of view, the use of these proteins in the footwear industry could lead to the application of high level technological innovation, but positive results could be clearly transferred to other industrial sectors in the European Union.

Acknowledgements

The authors acknowledge the Institute for Small and Medium Industry of the Generalitat Valenciana (IMPIVA) and the European Regional Development Fund (ERDF) for the partial financial support of this research project (IMDEEA/2011/103).

References

1   R.W. Smith, *Adhesives & Sealants Industry Magazine*, 2005, **12,** 5.
2   C. Saikumar, *J. Adhes. Sci. Technol.*, 2002, **16**, 543.
3   M. García-Gómez, J.A. del Ama and L. Artieda, *Arch. Prev. Riesgos Labor.*, 1998, **3**, 108.
4   H. Guo, F. Murray and S. Wilkinson, *J. Air & Waste Manage Assoc.*, 2000, **50**, 199.
5   Directive 2010/75/UE [OJ L 334, 17.12.2010, 17].
6   Directive 2008/98/CE [OJ L 312, 22.11.2008, 3].
7   Directive 98/24/CE [OJ L 131, 5.5.1995, 11].
8   F. Arán Aís, E. Orgilés Calpena, A.M. Torró Palau and C. Orgilés Barceló, *LederPiel*, 2005, 52.
9   Commission of the European Communities. A lead market initiative for Europe - COM (2007) 860 final. (European Commission, Brussels, 2007).
10 Commision of the European Communities. Europe 2020 – A strategy for smart, sustainable and inclusive growth – COM (2010) 2020 final. (European Commission, Brussels, 2010).
11 European Bioplastics. Bioplastics to pass the one billion tonne mark in 2011. Press release. Berlin, 12 May 2011. Available from: http://en.european-bioplastics.org/press/press-releases/.
12 K. Autumn, M. Sitti, Y.A. Liang, A.M. Peattie, W.R. Hansen, S. Sponberg, T.W. Kenny, R. Fearing, J.N. Israelachvili and R.J. Full, *Proc. Natl Acad. Sci. USA*, 2002, **99**, 12252.
13 K. Autumn, D. Santos, M. Spenko and M. Cutkosky, *J. Exp. Biol.*, 2007, **209**, 3569.
14 K. Autumn and A.M. Peattie, *Integr. Comp. Biol.*, 2002, **42**, 1081.
15 K. Autumn, A. Dittmore, D. Santos, M. Spenko, and M. Cutkosky, *J. Exp. Biol.*, 2006, **209**, 3569.
16 M. Varenberg, N.M. Pugno and S.N. Gorb, *Soft Matter*, 2010, **6**, 3269.
17 E. Artz, S. Gorb and R. Spolenak, *Proc. Natl Acad. Sci. USA*, 2003, **100**, 10603.
18 M.G. Langer, J.P. Ruppersberg and S. Gorb, *Proc. R. Soc. Lond.* B, 2004, **271**, 1109-2215.
19 W. Adlassing, T. Lendl, M. Perouka and I. Lang, "Deadly-Glue – Adhesive Traps of Carnivorous Plants" in *Biological Adhesive Systems – From Nature to Technical and Medical Application*, eds. J. von Byern and I. Grunwald, Springer-Verlag, Wien, 2010, 15-28.
20 Y. Zhou, S. Wu and V.P. Conticello, *Biomacromolecules*, 2001, **2**, 111.

21 F. Vollrath, *Rev. Mol. Biotechnol.*, 2000, **74**, 67.
22 O.Choresh, B. Bayarmagnai and R.V. Lewis, *Biomacromolecules*, 2009, **10**, 2852.
23 H. Zhao, C. Sun, RJ Steward and J.H. Waite, *J. Biol. Chem.*, 2005, **280**, 42938.
24 R.J. Steward, J.C. Weaver, D.E. Morse and J.H. Waite, *J. Exp. Biol.*, 2004, **207**, 4727.
25 S. DeMoor, J.H. Waite, M. Jangoux and P. Flammang, *Mar. Biotechnol.*, 2003, **5**, 45.
26 P. Flammang, A. Lambert, P. Bailly and E. Henebert, *J. Adhesion*, 2009, **85**, 447.
27 C.S. Wang and R.J. Stewart, *J. Exp. Biol.*, 2012, **215**, 351.
28 G. Li, C.S. Smith, Y.V. Brun and J.X. Tang, *J. Bacteriol.*, 2005, **187**, 257.
29 C.S. Smith, A. Hinz, D. Bodenmiller, D.E. Larson and Y.V: Brun, *J. Bacteriol.*, 2003, **185**, 1432.
30 P.H. Tsang, G. Li, Y.V. Brun, L. B. Freund and J.X. Tang, *Proc. Natl Acad. Sci. USA*, 2006, **103**, 5764.
31 E. Toh, H.D. Kurtz and Y.V: Brun, *J. Bacteriol.*, 2008, **190**, 7219.
32 G.G. Hardy, R.C. Allen, E. Toh, M. Long, P.J.B. Brown, J.L. Cole-Tobian and Y.V. Brun, 2010, *Mol. Microbiol.*, **76**, 409.
33 C.V. Benedict and J.H. Waite, *J. Morphol.*, 1986, **189**, 171.
34 J. A. Salerno and I. Goldberg, *Appl. Microbiol. Biotechnol.*, 1993, **39**, 221.
35 K. Inoue, Y. Takeuchi, S. Takeyama, E. Yamaha, F. Yamazaki, S. Odo and S. Harayama, *J. Mol. Evol.*, 1996, **43**, 348.
36 K. Inoue, Y. Takeuchi, D. Miki and S. Odo, *J. Biol. Chem.*, 1995, **270**, 6698.
37 R.Y. Floriolli, J. von Langen and J.H. Waite, *Mar. Biotechnol.*, 2000, **2**, 352.
38 C. Sun, J.M. Lucas and J.H. Waite, *Biomacromolecules*, 2002, **3**, 1240.
39 J.H. Waite and X. Qin, *Biochemistry*, 2001, **40**, 2887.
40 K. Inoue and S. Odo, *Biol. Bull.*, 1994, **186**, 349.
41 L.M. McDowell, L.A. Burzio, J.H. Waite and J. Schaefer, *J. Biol. Chem.*, 1999, **274**, 20293.
42 D.R. Flipula, S.M. Lee, R.P. Link, S.L. Strausberg and R.L. Strausberg, *Biotechnol. Prog.*, 1990, **6**, 171.
43 J.M. Lucas, E. Vaccaro and J.H. Waite, *J. Exp. Biol.*, 2002, **205**, 1807.
44 J. Yu, W.Wei, E. Danner, R.K. Ashley, J.N. Israelachvili, J.H. Waite, *Nat. Chem. Biol.*, 2011, **7**, 588
45 S.C.T Nicklisch, J.H. Waite, *Biofouling*, 2012, **28**, 865
46 M. Wiegemann, *Aquat. Sci.*, 2005, **67**, 166.
47 K. Kamino, K. Inoue, T. Maruyama, N. Takamatsu, S. Harayama and Y. Shizuri, *J. Biol. Chem.*, 2000, **275**, 27360.
48 M.T. Madigan, J.M. Martinko and J. Parker. "Brock: Biology of Microorganisms" Prentice-Hall International, Inc. 8edn., 1997.
49 G.A. O'Toole and R. Kolter, *Mol. Microbiol.*, 1998, **28**, 449.
50 S.M. Hinsa, M. Espinosa-Urgel, J.L. Ramos and G.A. O'Toole, *Mol. Microbiol.*, 2003, **49**, 905.
51 E. Deligianni, S. Pattison, D. Berrar, N.G. Ternan, R.W. Haylock, J.E. Moore, S.J. Elborn, J.S.G. Dooley, *BMC Microbiol.*, 2010, **10**:38
52 M.A. Davey and G.A. O'Toole, *Microbiol. Mol. Biol. Rev.*, 2000, **64**, 847.
53 L. Oligino and P. Fives-Taylor, *Infect. Immun.*, 1993, **61**, 1016.
54 P.E. Kolenbrander and R.N. Andersen, *Infect. Immun.*, 1990, **58**, 3064.
55 R.N. Andersen, N. Ganeshkumar and P.E. Kolenbrander, *Infect. Immun.*, 1993, **61**, 981.
56 J.S. Sampson, S.P. O'Connor, A.R. Stinson, J.A. Tharpe and H. Russell, *Infect. Immun.*, 1994, **62**, 319.
57 N. Ganeshkumar, M. Song and B.C. McBride, *Infect. Immun.*, 1988, **56**, 1150.

58   F.F. Correia, J.M. DiRienzo, T.L. McKay and B. Rosan, *Infect. Immun.*, 1996, **64**, 2114.

59   N. Ganeshkumar, P.M. Hannam, P.E. Kolenbrander and B.C. McBride, *Infect. Immun.*, 1991, **59**, 1093.

60   M.K. Yeung, *Crit. Rev. Oral Biol. Med.*, 1999, **10**, 120.

61   S. Hamada, A. Amano, S. Kimura, I. Nakagawa, S. Kawabata and I. Morisaki, *Oral Microbiol. Immunol.*, 1998, **13**, 129.

62   J.O. Cisar, E.L. Barsumian, R.P. Siraganian, W.B. Clark, M.K. Yeung, S.D. Hsu, SD. Curl, A.E. Vatter and A.L. Sandberg, *J. Gen. Microbiol.*, 1991, **137**, 1971.

63   M.K. Yeung and J.O. Cisar, *J. Bacteriol.*, 1990, **172**, 2462.

64   M.K. Yeung and J.O. Cisar, *J. Bacteriol.*, 1988, **170**, 3803.

65   M.J. Escoto, M.R. Rodríguez, M.M. Sánchez, R.P. Mellado, "Biodegradable adhesives" in *Recent advances in research on biodegradable polymers and sustainable composites*, eds. A. Jimenez, G.E. Zaikov, Nova Science Publishers, New York, 2007, Vol. 2, Chapter 8, pp. 89-100

66   A. Kotiranta, M. Haapasalo, K. Kari, E. Kerosuo, I. Olsen, T. Sorsa, J.H. Meurman and K. Lounatmaa, *Infect. Immun.*, 1998, **66**, 4895.

67   A. Kotiranta, K. Lounatmaa and M. Haapasalo, *Microbes and Infection*, 2000, **2**, 189.

68   P. Lechant, L. Hummel, S. Rousseau and I. Moszer, *Nucleic Acids Res.*, 2008, **36**, D469

69   S.F. Altschul, T.L. Madden, A.A. Schaffer, J. Zhang, Z. Zhang, W. Miller and D.J. Lipman, *Nucleic Acids Res.*, 1997, **25**, 3389

70   J.C. Wootton, E. M. Gertz, R. Agarwala, A. Morgulis, A.A. Schaffer and Y. Yu, *FEBS J.*, 2005, **272**, 5101

71   N.R. Stanley and B.A. Lazarezza, *Mol. Microbiol.*, 2004, **54**, 917.

72   E. Shimoni, U. Ravid and Y Shoham, *J. Biotechnol.*, 2000, **78**, 1.

73   M.A. Titok, J. Chapuis, Y.V. Selezneva, A.V. Lagodich, V.A. Prokulevich, S.D. Ehrilich and L. Janière, *Plasmid*, 2003, **49**, 53.

74   C. Ruiz, F.I. Pastor and P. Díaz, *Lett. Appl. Microbiol.*, 2005, **40**, 218.

75   J.K. Kim, K.J. Park, K.S. Cho, S.W. Nam, T.J. Park and R. Bajpai, *Bioresour. Technol.*, 2005, **96**, 1897.

76   K. Kobayashi et al., *Proc. Natl Acad. Sci. USA*, 2003, **100**, 4678.

77   F. Kunst et al., *Nature*, 1997, **390**, 249.

78   V. Barbe, S. Cruveiller, F. Kunst, P. Lenoble, G. Meurice, A. Sekowska, D. Vallenet, T. Wang, I. Moszer, C. Médigue and A. Canchin, *Microbiology*, 2009, **155**, 1758

79   M. Morikawa, *J. Biosc. Bioeng.*, 2006, **101**, 1.

80   J.C. Zweers, I. Barák, D. Becher, A.J.M. Driessenet, M. Hecker, V.P Kontinen, M.J. Saller, L. Vavrorá and J.M. van Dijl, *Microb. Cell Fact.*, 2008, **7**, 10.

81   A. del Campo and E. Arzt, *Chem. Rev.*, 2008, **108**, 911.

82   H. Lee, B.P. Lee and P.B. Messersmith, *Nature*, 2007, **448**, 338.

83   L. Ninan, J. Monahan, R.L. Stroshine, J.J. Wilker and R. Shi, *Biomaterials*, 2003, **24**, 4091.

84   J. Wang, C. Liu, X. Lu and M. Yin, *Biomaterials*, 2007, **28**, 3456.

85   L. Ninan, R.L. Stroshine, J.J. Wilker and R. Shi, 2007, *Acta Biomater.*, **3**, 687.

86   B.P. Lee, P.B. Messersmith, J.N. Israelchvili and J.H. Waite, *Annu. Rev. Mater. Res.*, 2011, **41**, 99.

87   L.A. Burzio and J.H. Waite, *Protein Sci.*, 2001, **10**, 735.

88   C. Fant, J. Hedlund, F. Höök, M. Berglin, E. Fridell, H. Elwing, *J Adhesion*, 2010, **86**, 25.

89 J. Hedlund, M. Anderson, C. Fant, R. Bitton, H. Bianco-Peled, H. Elwing, M. Berglin, *Biomacromolecules*, 2010, **10**, 845.

90 J.R. Burkett, J.L. Wojtas, J.L. Cloud, J.J. Wilker, *J Adhesion*, 2009, **85**, 601.

91 H.G. Silverman and F.F. Roberto, *US Pat.*, 6,987,170.

92 H.G. Silverman and F.F. Roberto, *US Pat.*, 6,995,012.

93 H.G. Silverman and F.F. Roberto, 2007, *Mar. Biotechnol.*, **9**, 661.

94 C.V. Benedict and P.T. Picciano, *European Patent* EP0243818B1.

95 K.J. Maugh, *Canadian Patent* CA 1282020

96 J. Gómez. Master in science, technology and society ESST 1995-1996. Universidad Autónoma de Madrid-IADE (1997).

97 Directive 2000/54/EC [OJ L 262, 17.10.2000, p. 21.]

98 N. Cuesta, M.J. Escoto, F. Arán and C. Orgilés, *J. Adhesion*, 2012, **88**, 294.

99 M.A. Hamon and B.A. Lazazzera, *Mol. Microbiol.*, 2001, **42**, 1199.

100 N. Cuesta-Garrote, M.J. Escoto-Palacios, C. Orgilés-Barceló, "Proteínas para la formulación de adhesivos bioinspirados" in *Tendencias en Adhesión y Adhesivos. Volumen IV*, eds. S. Flórez, C. Jiménez, B. Pérez and R. Rodríguez, San Sebastián 2011, pp. 20-31

SURFACE MODIFICATION FOR OPTIMAL BONDING/DEBONDING

ENDOTHELIAL CELLS ADHESION ON MODIFIED POLYURETHANE SURFACE AS THE WAY TO FABRICATE A NOVEL MATERIAL FOR CARDIOSURGERY

P.A. Zietek*, B.A. Butruk and T. Ciach

Faculty of Chemical and Process Engineering, Warsaw University of Technology, Poland
*P.Zietek@ichip.pw.edu.pl

1   INTRODUCTION

1.1. Different Approaches to Heart Tissue Regeneration

Since heart diseases have become the number one cause of death worldwide,[1] there is a strong interest in searching of efficient methods of heart tissue regeneration. The possible way of therapy depends on the degree of tissue damage. In case of i.e. nonischemic cardiomyopathy there is the opportunity of applying the left ventricular assist device (LVAD) therapy. LVAD-induced mechanical unlock of a heart would lead to a complete myocardium recovery and a patient may not need any additional treatment.[2,3] However, in case of more serious heart diseases, it is obligatory for a patient to undergo a replacement procedure of a whole heart or its parts.

In order to avoid such an extensive and dangerous surgery, for more than fifty years attention has being paid for tissue engineering methods of heart tissue regeneration.[4] These include an injured heart support by introduction of cell-seeded or cell-free materials into the heart. The function of cell-free materials is to replace a dilated, soft myocardium and thus improve ventricular function[5-7] and then can be repopulated with endogenous cells.[8] Another approach comprises the use of pre-formed three-dimensional scaffolds of synthetic and/or natural polymers that would aid in the introduction of the cells to the heart. In the *in vitro* tissue formation method, living tissue implants are generated outside the human body.[9,10] In situ tissue engineering includes applying materials presenting active biomolecules. These allow controlling and guiding endogenous cell repopulation and subsequent tissue formation inside the human body by using the intrinsic regenerative capacity of the body.[11]

1.2 Artificial Heart as an Alternative Approach

The main problem concerning approaches mentioned above are the weak mechanical properties of fabricated implants.[12] Hence, replacement of the damaged heart with an artificial heart made of synthetic material seems to be an interesting point of view. However, there are many expectations that have to be met to minimize harmful effects that could arise.

Because of its destination, an artificial prosthesis would directly and constantly contact blood for a long time – if possible the lifetime of the patient. An implant surface becomes a border between living body and artificial live-rescuing device. Although implantable materials are carefully chosen, they still do not interact with blood components as natural vessel tissue does. That results in a number of unwanted phenomena occurring on the prosthesis surface. These effects form a complex biochemical process that has a great influence on a patient's health, even long after implantation. Hence, artificial heart has to be made from a material that should cooperate with blood without any side-effects. The only way to achieve this goal is to cover artificial prosthesis's material with endothelial cells that form a monolayer in the inner part of blood vessels and heart. This approach considers *in vitro* cell pre-seeding and cultivation before implantation. That is the subject of the present work.

Below, we describe the advantages of PU in terms of artificial heart construction, the necessity of covering the surface of a PU implant with endothelial cells and the method of *in vitro* introducing endothelial cells onto PU.

1.3 Physiological Effects Caused by Artificial Heart Prosthesis

The very first event that happens on the surface of artificial material is the adsorption of water and small ions together with small proteins, such as kininogen or fibrinogen. These are components that can trigger the blood coagulation cascade.[13] This is a rapid process that requires just traces of initializing factors and results in the formation of blood clots. In case of artificial heart, vessels or any other artificial implant it is strongly unwanted, because clots are likely to detach from the surface and cause thrombosis that can lead to a stroke.[14] Moreover, the presence of proteins increases the hydrophilicity of implanted material surface and so blood cells start to attach. Monocyte adhesion can cause activation of extensive inflammation response: activation of neutrophils and leukocytes and later also macrophages.[15] Porosity of the surface increases as well. That stimulates adhesion of platelets and other factors that eventually leads to formation of blood clots.[16]

Since artificial heart should serve well for the whole patient's life, it is crucial that a heart prosthesis's material would not trigger any of these harmful effects.

1.4 Artificial Heart Prosthesis Made from PU

Nowadays PUs are the most promising polymers used in fabrication of artificial heart prosthesis. Therefore, PUs were chosen to be the focus of the present work.

PUs are segmented copolymers that constitute a diverse group of chemical compounds. The selection of monomers used in the PU synthesis can produce materials with different mechanical characteristics, which makes PU an attractive biomaterial. Also, among all polymers, PUs are characterized by the best hemocompatibility (compatibility with blood).[17]

Hemocompatibility of PUs is satisfying but not impeccable. Also, *in vivo* studies indicated degradation of unprotected PU by surface oxidation,[18,19] hydrolysis[20] and radical processes caused by blood cells metabolism.[21] Usually, a polymer such as PU undergoes further chemical or physical modification. A number of methods have been applied: hydrophilization[22] or hydrophobization[23,24] of the polymer surface and immobilization of biomolecules such as phospholipids[25] or heparin.[20] None of them was completely successful. The most advanced method is to cultivate endothelial cells on previously prepared polymer surface, which is described below.

1. Intima lined by endothelial cells
2. Internal elastic lamina
3. Media containing sooth muscle cells
4. External elastic lamina
5. Adventitia with vasa vascorum & nervi vascorum

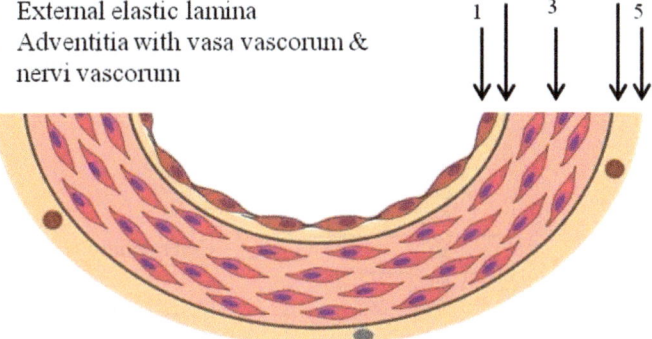

Figure 1 *Transverse section through the wall of a blood vessel showing its different constituents including the inner endothelial monolayer.*[26]

1.5. Role of Endothelial Cells

Endothelial cells form a monolayer that covers the inner part of natural blood vessels and heart. This tight monolayer forms a barrier between blood and smooth muscle cells (Figure 1). Their strategic localization places endothelial cells in very first line of contact with cells and substances migrating from blood to tissue. The endothelium tissue constantly controls blood cells activity since it produces a large variety of both activating and inhibiting factors.[27] Endothelium, via a large number of mechanisms not yet known in-depth, plays crucial role in regulation of vasomotorics,[28,29] hemostasis[27,30,31] and angiogenesis.[32]

1.6 Adhesion of Endothelial Cells

Adhesion of endothelial cells on a vessel surface leads to formation of tight monolayer of cells. The process is based on interactions between endothelial transmembrane receptors and ligands that are specific molecules bonded by receptors. Ligands are found in the extracellular matrix (ECM) that *in vivo* composes a base for endothelial cells. ECM includes peptides, proteins, peptidoglycans and glycoproteins.[33] All of these play a role in regulating endothelial cells metabolism.[33] Receptor-ligand binding starts a cascade of biochemical signals inside a cell that finally results in cell attachment to a surface on which a ligand is present.

1.6.1 Adhesion Receptors - Integrins and their Ligands. Adhesion receptors have already been widely described in the literature.[35] Generally, they are divided into four groups: integrin, cadherin, Ig-CAM, and selectin family.[36] Endothelial cell adhesion is mostly dependent on integrins. Each integrin is a heterodimer that contains an α and a β subunit with each subunit having a large extracellular domain, a single membrane-spanning region, and in most cases a short cytoplasmic domain (Figure 2).[34]

Typically, integrins mediate binding to large ECM proteins, such as collagen, laminin, vitronectin and fibronectin.[33,37] Thanks to the development of analytic methods, it was discovered that all ligands dedicated to cell adhesion receptors contain specific amino

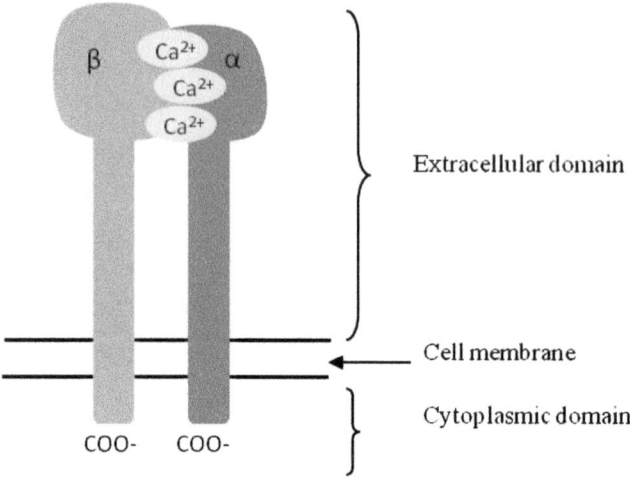

Figure 2   *A Schematic drawing showing the structure of integrins: α and β subunit, each including a large extracellular domain, a single membrane-spanning region and a short cytoplasmic domain.*[34]

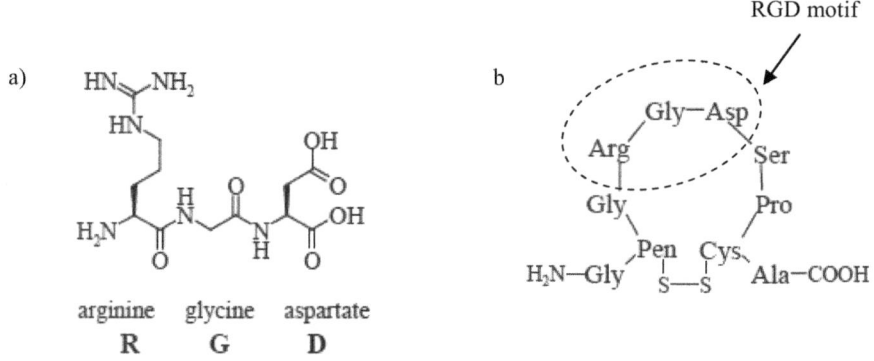

Figure 3   *a) Structure of RGD tripeptide included in adhesion receptors ligands, b) Structure of a cyclic peptide with RGD sequence.*

acid sequence that determines a proper interaction with a receptor. The sequence that is most commonly mentioned in literature is a RGD[33,38-40] tripeptide (Figure 3a). It has been found e.g. in fibronectin and vitronectin.[38]

It has also been proven that stereochemistry of an adhesion sequence affects cell attachment activity. Cyclopeptides containing RGD sequence (Figure 3b) are more affine to integrins than corresponding linear ones.[38,41-43] Apart from RGD, there are also many different adhesion sequences known, such as LDV,[44] REDV,[45,46] YIGSR,[47] and PDSGR.[33,38]

1.6.2 Effects of Integrin-ligand Interaction.   Figure 4 illustrates the idea of endothelial cells attachment on a surface rich with adhesion ligands. Binding to integrin receptors activates various signaling paths that, among other functions, mediate cell

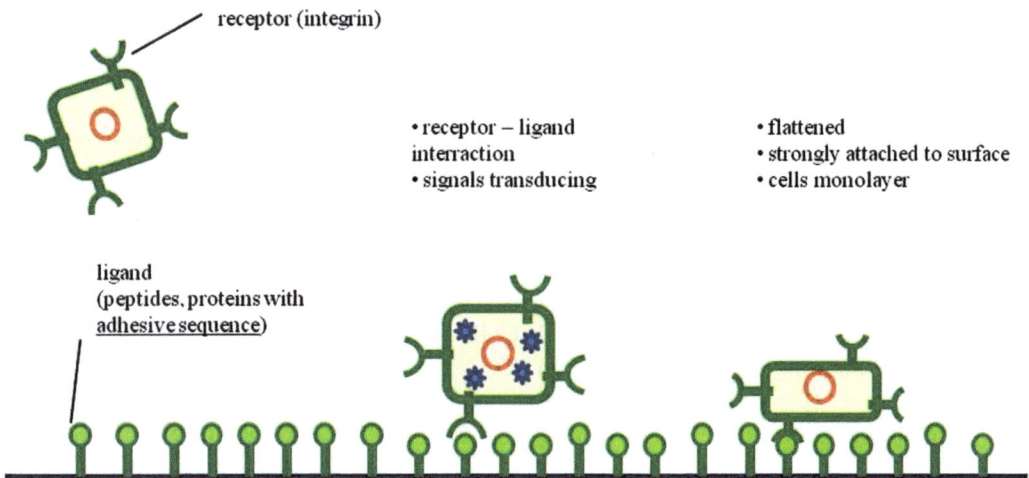

receptor (integrin)

• receptor – ligand
interraction
• signals transducing

• flattened
• strongly attached to surface
• cells monolayer

ligand
(peptides, proteins with
adhesive sequence)

Figure 4   *Model of endothelial cells attachment on a surface presenting adhesion ligands.*

attachment, proliferation, differentiation and organization of the actin cytoskeleton.[33,37,48] Thus, a change of cell shape is a visible symptom of proper attachment and is a result of biochemical processes initiated by ligand binding. Cells become more flattened and able to form a network of focal adhesions to a surface.[48] Finally, a cell is strongly anchored to a surface thanks to integrins and other adhesive membrane molecules (e.g. lipids and syndecans) and capable of communicating with other cells next to it.[44, 49]

   1.6.3   Conditions of Proper Endothelial Cell Anchorage to a Surface.   Receptor-ligand interaction requires free space around the two components so that both of them can take the right conformation.[39] Strong covalent bond between a ligand and a surface is very important. Also, any unbounded peptides or proteins should be removed. Endothelial cells that are deprived of anchorage to a ligand die by an apoptotic signal.[50,51]

1.7.   Aim of Work

The goal of this work was to develop the repeatable and easy to proceed method of fabricating a material that could construct the inner side of an artificial heart. The final material would comprise PU coated with endothelial cells *in vitro*. Thus, the material would present suitable mechanical properties and interact with blood in the way natural tissue does.
   PU surface was grafted with collagen that is a ligand for endothelial cell adhesion mediated by integrins. Then human endothelial cells were cultivated on the material. The present work includes a detailed description of the collagen binding to the PU surface and results from the cell seeding. The main challenge was to obtain a stable junction between collagen and PU surface so endothelial cells would properly interact with it. Also, we focused on the elimination of any factors that could interfere with a correct cell-material interaction, such as surface roughness or improper collagen conformation.

2 MATERIALS AND METHODS

2.1   Materials

For the PU film preparation we used biomedical PU Estane 5715 P in form of grains (Lubrizol) and THF (Fluka), 30% (v/v), as the solvent. During the PU modification process we used: H_2O_2 (Carlo Erba), 65% (v/v) HNO_3 (Lach-Ner s.r.o), $(NH_4)_4Ce(SO_3)_4$ (Riedel-de-Haën), $CH_2CHCOOH$ (Fluka), EDC (Sigma Aldrich), NHS (Sigma Aldrich), soluble collagen in citric buffer, average molecular mass 340kDa (Proteina). For –COOH groups analysis TBO (Sigma Aldrich), NaOH (Carlo Erba) and were applied. For endothelial cells culture, HUVEC (Lonza) medium was used. Cells were stained with the use of EBM-2 (Lonza) Double Staining Kit (Sigma Aldrich).

2.2   Methods

In this work PU underwent superficial modification. The aim was to enrich its surface with collagen molecules since it is a compound of ECM and was reported to interact with integrins.[52] PU grafted with collagen was used in endothelial cells culture to check whether cells would anchor to the modified surface.

Surface modification was accomplished by a three-step chemical procedure. Figure 5 schematically shows PU surface after modification. Our assumption was that only surface undergoes a modification process, therefore mechanical properties provided by PU would remain unaffected. Applied conditions were as mild as possible, because the material must not be toxic or harmful for the cells. All experiments and measurements were carried out once.

2.2.1   Preparation of PU Films. A 20% (w/v) solution of PU in THF was prepared. It was stirred for 24h to dissolve the polymer and then left to rest for another 24h. Then the solution was poured on a glass via a film applicator. The PU film was air-dried for 3 days and next in an oven at 40°C for another 3 days. Obtained PU films were 0.3mm in thickness.

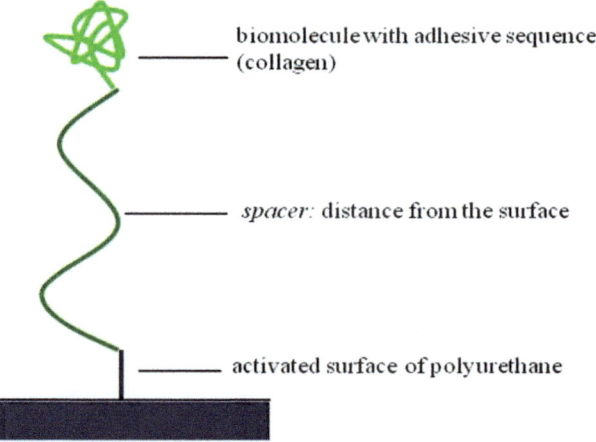

Figure 5.   *PU surface after modification shown schematically. Distinct elements that were introduced at each step of the modification process are pointed out.*

2.2.2 Activation of PU Surface. PU is chemically inert - it does not contain any superficial reactive chemical group. Thus, the very first step of modification is to activate PU surface. In the present work a photooxidation method was applied.[53-55] Figure 6 presents a scheme of this step. Photooxidation is a popular method allowing the introduction of –OOH and –OH groups on the polymer surface. It does not require complex equipment. Also, radicals produced by this method are very reactive, so the reaction yield is high.

Pieces of PU foil were put into flat containers filled with small amount of a 30% (v/v) solution of H_2O_2. Containers were then put inside a UV lamp chamber. Samples were exposed to UV-C radiation for a defined period of time. Different times of exposure were tested – form 1h to 12h. After that, samples were washed three times with distilled water and dried in an oven at 40°C. Samples were analyzed with the use of FTIR-ATR spectroscopy. A Thermo Scientific Nicolet™ 6700 spectrometer was used for obtaining the spectra, and OMNIC 8 software was used for spectrum analysis.

Figure 6. *A scheme of the activation process of PU surface. The exposure to UV-C radiation leads to superficial –OH groups appearing.*

2.2.3 Spacer Molecules Grafting. As it was mentioned in point 1.4.4, it is essential that there is some free space provided between a ligand and cell receptor. Thus, it is recommended not to couple ligands directly to superficial reactive groups, but to do it indirectly through long *spacer* molecules grafting. At this stage, *in situ* radical polymerization of acrylic acid initiated by ceric ions was proposed.[56] The process is presented in Figure 7.

Activated PU samples were placed in a water solution containing 1.5% (v/v) HNO_3. The solution was heated up to a defined temperature. Then 0.1% (w/v) $(NH_4)_4Ce(SO_3)_4$ and a particular amount of $CH_2CHCOOH$ were added and stirred. Reaction was carried out

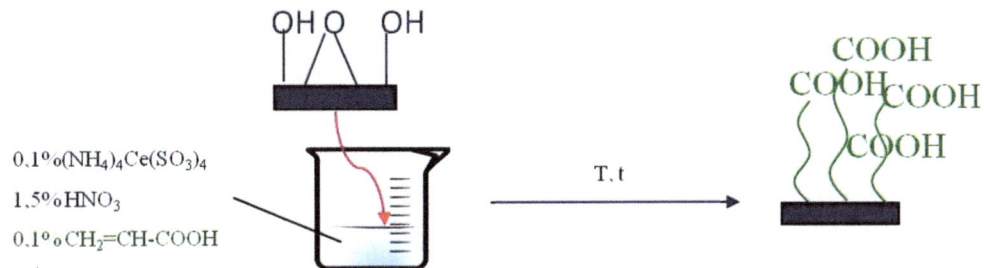

Figure 7. *A scheme of the spacer molecules grafting process. Hydroxyl groups present on the activated PU surface allow poly(acrylic acid) chains grafting through in situ polymerization initiated by ceric ion reduction.*

with constant stirring for particular period of time. Parameters applied were: 25, 35, 40 and 45°C, 0.5, 1.5 and 2.5h of reaction and 5, 10, 15 and 20% (v/v) of CH₂CHCOOH. After the predefined time of reaction, the samples were taken out from the solution, washed three times with distilled water with stirring and dried in an oven at 40°C. The long chains of polyacrylic acid grafted to the PU surface enrich it with carboxyl groups, thus these groups were used for evaluation. The amount of superficial -COOH groups was evaluated colorimetrically with the use of TBO[57] (Figure 8), one mole of TBO binding to one mole of –COOH. The absorbance of TBO washing solutions (i.e. acetic acid) was measured at λ=630nm with a Helios γ spectrophotometer (by Thermo Electron Corporation). Concentrations of TBO that correspond to concentrations of -COOH were estimated from a calibration curve.

2.2.4 Collagen Immobilization.   Before this stage was carried out, samples obtained with the use of best parameters were chosen.

For coupling collagen to carboxylated PU surface, a 2-step carbodiimide method[58] was applied. Figure 9 presents a scheme of this method. Carboxyled PU foils were placed in flat containers filled with an aqueous solution of pH=6 containing 2% (w/v) EDC and 2% (w/v) NHS. After 15minutes samples were taken out and washed with a phosphate buffer solution of pH=6 and carefully dried with paper towel. After that samples were put into an aqueous bovine collagen (mixture of type I and type III) solution. Collagen was used in five-fold molar excess with reference to calculated –COOH groups. Reaction was carried out for 2h. Samples were then washed three times with distilled water and kept in PBS solution.

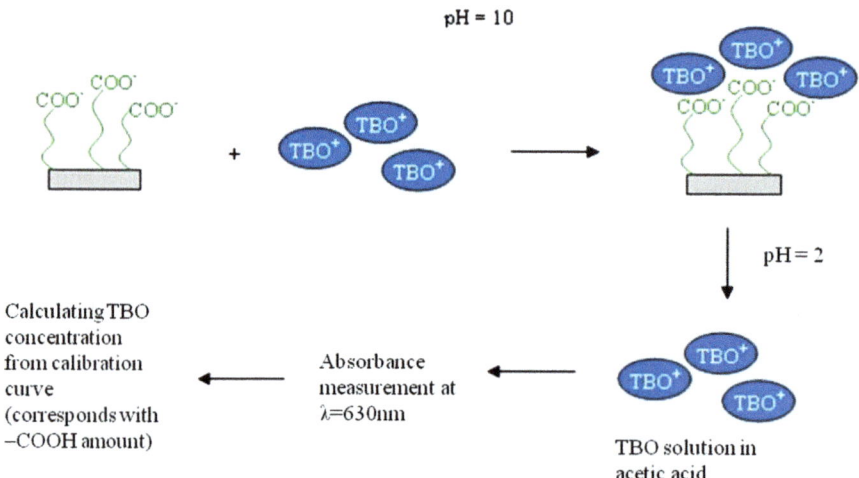

Figure 8   *A scheme of the colorimetric method for the evaluation of –COOH group concentration.[57] TBO binds to carboxyl groups that are present on PU surface after spacer grafting. After decoupling, the absorbance of the washing solution is measured. The amount of –COOH groups is estimated from a calibration curve.*

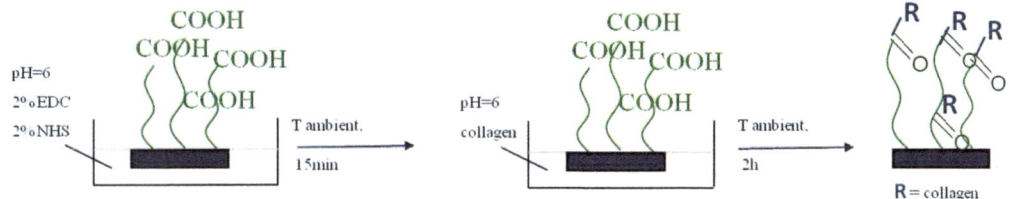

Figure 9 *An illustration of the collagen immobilizing. Amine groups from collagen react with –COOH groups from PU surface. As a result, a peptide bond is formed that provides a strong junction between PU surface and collagen.*

2.2.5 Endothelial Cell Culture. Samples of unmodified PU and PU with immobilized collagen were sterilized with UV radiation for 30min. Cells culture was carried out using HUVEC in standard medium for endothelium for 2 days at 37°C. In order to assess cell viability, cells were stained with Double Staining Kit.[59] Samples were observed every day during cultivation using a transverse optical fluorescence microscope Nikon Eclipse Ti-U.

3 RESULTS AND DISCUSSION

3.1 Activation of PU Surface

Figure 10 shows a part of FTIR-ATR spectra obtained for unmodified PU and PU samples activated for 1 to 5h. Samples exposed to UV-C radiation for 12h were thermically damaged because of the heat emitted by the UV lamp, thus they were not analyzed.

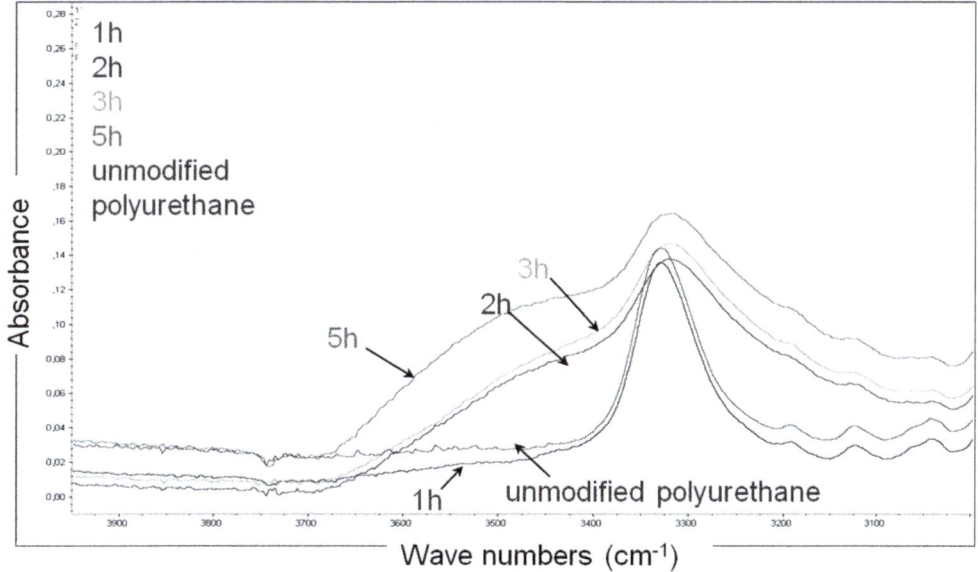

Figure 10 *A part of FTIR-ATR spectra of unmodified PU and PU activated for 1 to 5 hours. A characteristic signal at 3600-3000 cm⁻¹ range that comes from –OH groups can be observed.*

There is a characteristic signal appearing at 3600-3000 cm^{-1}range that comes from –O-H bonds. The peak is clearly increasing and becomes the strongest for a sample activated for 5h. Hence, 5 hours of exposure was accepted as the most efficient. As for PU surface hydroxylation via UV radiation, other authors propose similar exposure times, e.g. 4h.[56]

3.2 *Spacer* Molecules Grafting

Figure 11a-d shows charts presenting the amount of –COOH groups per 1cm^2 of sample in function of time for different CH$_2$CHCOOH concentrations and different reaction times. Charts show that carrying out acrylic acid grafting at higher temperatures (40 and 45°C) does not allow to achieve higher amounts of superficial carboxyl groups. Also, –COOH concentration is roughly constant during the whole process. It is probably caused by a side reaction – homopolymerization of acrylic acid in solution – which rate increases at higher temperatures. Homopolymerization plays a main role also for higher concentrations of acrylic acid. That results in low yields of grafting (Figure 11d). Homopolymerization as a result of high concentration of acrylic acid and high temperature during the process was also reported by other authors.[60] For 5, 10 and 15% of acrylic acid and temperatures of 25 and 35°C results were similar. However, only samples prepared with the use of 5% of acrylic acid presented an acceptably smooth surface, without any roughness and clots. The largest amount of superficial -COOH groups (3.28 mM -COOH/cm^2) was obtained for 5% (v/v) of acrylic acid, 35°C and 1.5h (Figure 11a). Therefore, those parameters were chosen as the most suitable.

It was previously reported that the increase of reaction temperature leads to high acrylic acid grafting yield.[60] However, unlike in the present work, it has been shown that the grafting yield grows with the temperature increase up to 40°C (cellulose fiber) or 50°C (PET fiber).[60,61] Nevertheless, the roughness of samples after grafting was not mentioned, probably because it was not significant in those researches. In the present work, the smoothness of surface is crucial.

A further increase in temperature decreased the graft yield. This was also indicated in other research.[61] At higher temperature, the collision between monomer free radicals also increased, wherein the homopolymer formation becomes predominant.

As it was shown in the figure 11, reaction time plays an important role in acrylic acid grafting. The maximum amount of –COOH appears at 1.5h, and when the reaction proceeds further, a decrease can be observed. In the literature, authors mention different times when the largest –COOH amount is achieved (from 1h[60,61] to 2-3h[62]), yet the decrease is reported in all these works. The explanation may be that the TBO method applied to estimate the amount of –COOH groups has its limitations. The complexation of TBO may be hindered by the presence of the grafted chains when the graft yield is high.[61]

3.3 Endothelial Cell Culture

Table 1 shows parameters applied to prepare samples used for endothelial cell culture. Figure 12 presents pictures from optical microscope (magnification 100x) of HUVEC cultivated on unmodified PU and PU with immobilized collagen. Obtained pictures indicated that for HUVEC deprived of anchorage on unmodified PU surface, after one day of cultivation only red cells were observed. Hence there were no living cells on unmodified PU after one day of cultivation. However, on the surface of PU enriched with collagen the cells survived (Figure 12b). Furthermore, after 2 days of cultivation, cells significantly

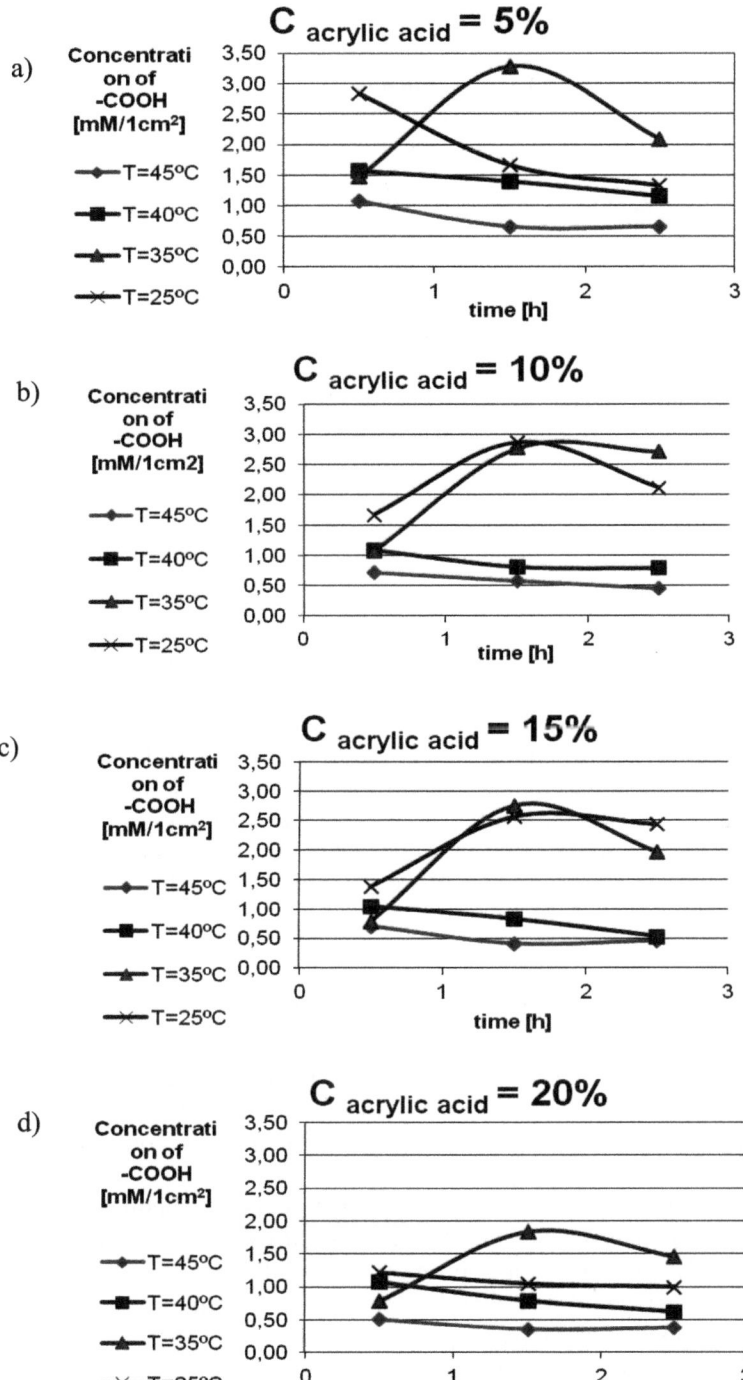

Figure 11 *Charts presenting amount of –COOH groups per 1cm² of sample in function of time for different CH₂CHCOOH concentrations and different reaction times (TBO method).*

Table 1 *Parameters applied to prepare samples used for endothelial cell culture*

Step	Parameters
Activation of PU surface	5h of exposure to UV-C radiation
Spacer molecules grafting	5% (v/v) of acrylic acid 35°C 1.5h
Collagen immobilization	5:1 molar ratio of collagen to -COOH groups

changed their morphology and seemed to be strongly attached to the surface (Figure 12c). This is a proof of integrin-mediated cell adhesion via ligand (collagen) binding. Besides, there are characteristic twin cells visible (Figure 12c, circled) that most likely were formed after cell division. This also indicates the proper anchorage via integrins.

The change in the cell shape was expected as the positive premise of the cell anchorage to the surface via integrin-mediated mechanism.[63-66] However, the density of HUVEC should be higher. It was shown that it is possible to obtain a tight monolayer of HUVEC on a PU grafted with gelatin[64] and on modified titanium surface.[67] Besides, obtaining a tight cell coating is crucial because of thrombogenicity of collagen. Collagen molecules exposed on the surface of an implant would cause a strong inflammatory response starting with platelet adhesion.[65]

We observed cells proliferation 2 days after seeding which has been also previously noted. Sgarioto et al. reported that HUVEC proliferate most intensively 2 days after seeding on materials coated with collagen and other ECM proteins.[68] Nevertheless, in the present research we did not succeed in longer HUVEC cultivation, whereas other authors managed to obtain viable cells after 7 days of cultivation.[65] This was most likely due to collagen degradation by the cells enzymes.

4 CONCLUSIONS AND FUTURE PROSPECTS

The aim of the present work was to propose a method to fabricate a material that is composed of PU coated with HUVEC. In the considered approach, the patient's endogenous cells would be pre-seeded and pre-cultivated *in vitro* outside the body until the cells form a tight, healthy monolayer.

It was demonstrated that the process of three-step chemical modification was successful. Samples analysis showed effectiveness of each step. Final materials were not toxic for endothelial cells and clearly promoted cells adhesion, proliferation and change of cell shape. The advantage of the method is that it does not require any complex equipment and thus can be applied in any laboratory and be utilized on a large scale. Besides, it is free from organic solvents, so there is no need of special preparation before cell seeding. The proposed method of endothelialization is promising and can be applied to other base polymers and proteins or peptides.

Since the research was promising, we are going to apply this method to graft short peptides such as RGD or YIGSR onto PU. That may lead to obtain a monolayer of endothelial cells that would completely separate the PU surface from blood. After the tight cell coating would be achieved, the material should be tested by platelet-rich plasma (PRP) incubation to examine platelets activation.

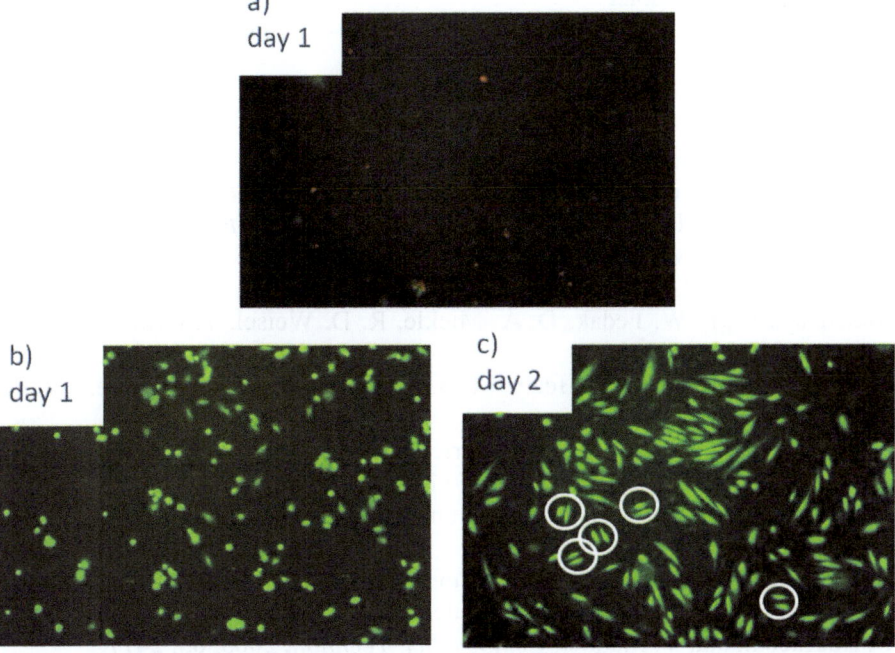

Figure 12   *Pictures from optical microscope (magnification 100x) showing HUVEC on a) unmodified PU after 1 day of cultivation, b) PU with immobilized collagen after 1 day of cultivation and c) PU with immobilized collagen after 2 days of cultivation. Cell pairs formed after cell division are circled.*

Abbreviations

CH$_2$CHCOOH	acrylic acid
EBM-2	Endothelial Basal Medium-2
ECM	extracellular matrix
EDC	1-ethyl-3-(3-dimethylaminopropyl) carbodiimide
FTIR-ATR	Fourier-transform infrared spectroscopy - attenuated total reflectance
HUVEC	human umbilical vein endothelial cells
Ig-CAM	immunoglobulin cell adhesion molecules
LDV	leucine-aspartate-valine
LVAD	left ventricular assist device
NHS	N-hydroxysuccinimide
PBS	phosphate-buffered saline
PDSGR	proline-aspartate-serine-glycine-arginine.
PRP	platelet-rich plasma
PU	polyurethane
REDV	arginine-glutamate-aspartate-valine
RGD	arginine-glycine-aspartate
TBO	Toluidine Blue O
THF	tetrahydrofuran

UV   ultraviolet
YIGSR   tyrosine-isoleucine-glycine-serine-arginine

References

1. WHO, *Fact sheet N°317 - Cardiovasular diseases*, 2009.
2. M. Ibrahim, C. Terracciano and M. H. Yacoub, *Curr. Cardiol. Rep.*, 2012, **14**, 392.
3. M. Guglin and L. Miller, *Curr. Treat. Options Cardiovasc. Med.*, 2012, **14**, 370.
4. C. E. Schmidt and J. M. Baier, 2000, **12**, 2215-2231.
5. W. Dai, L. E. Wold, J. S. Dow and R. A. Kloner, *J. Am. Coll. Cardiol.*, 2005, **46**, 714.
6. K. Matsubayashi, P. W. Fedak, D. A. Mickle, R. D. Weisel, T. Ozawa and R. K. Li, *Circulation*, 2003, **108**, 11219.
7. S. T. Wall, J. C. Walker, K. E. Healy, M. B. Ratcliffe and J. M. Guccione, *Circulation*, **114**, 2627.
8. C. V. C. Bouten, P. Y. W. Dankers, A. Driessen-Mol, S. Pedron, A. M. A. Brizard and F. P. T. Baaijens, *Adv. Drug Del. Rev.*, 2011, **63**, 221.
9. M. N. Giraud, C. Armbruster, T. Carrel and H. T. Tevaearai, *Tissue Eng.*, 2007, **13**, 1825.
10. J. Leor, N. Landa and S. Cohen, *Expert Rev. Cardiovasc. Ther.*, 2006, **4**, 239.
11. M. Avci-Adali, G. Ziemer and H. P. Wendel, *Biotechnol. Adv.*, 2010, **28**, 119.
12. A. R. Boccaccini and V. Maquet, *Compos. Sci. Technol.*, 2003, **63**, 2417.
13. J. M. Berg, J. L. Tymoczko and L. Stryer, eds., *Biochemia*, Wydawnictwo Naukowe PWN, Warszawa, 2005.
14. J. A. Bittl, *J. Am. Coll. Cardiol.*, 1996, **28**, 368.
15. H. M. van Beusekom, W. J. van der Giessen, R. van Suylen, E. Bos, F. T. Bosman and P. W. Serruys, *J. Am. Coll. Cardiol.*, 1993, **21**, 45.
16. P. Turbill, T. Beugeling and A. A. Poot, *Biomaterials*, 1996, **17**, 1279.
17. A. Tiwari, H. Salacinski, A. M. Seifalian and G. Hamilton, *Cardiovasc. Surg.*, 2002, **10**, 191.
18. M. Szycher and W. A. McArthur, *Corrosion and Degradation of Implant Materials: Second Symposium*, American Society for Testing and Materials, Philadelphia, 1985, 308-321.
19. Y. Wu, C. Sellitti, J. M. Anderson, A. Hiltner, G. A. Lodoen and C. R. Payet, *Journal of Applied Polymer Science*, 1992, **46**, 201.
20. I. Resiak and G. Rokicki, *Polimery*, 2000, **45**, 589.
21. S. K. Phua, E. Castillo, J. M. Anderson and A. Hiltner, *J. Biomed. Mater. Res.*, 1987, **21**, 231.
22. M. Trzaskowski, B. Butruk and T. Ciach, *Mater. Sci. Eng., C*, 2012, **32**, 1601.
23. P. Zietek, B. Butruk and T. Ciach, *Central European Journal of Chemistry*, 2011, **9**, 1039.
24. M. Dabagh, M. J. Abdekhodaie and M. T. Khorasani, *J. Appl. Polym. Sci.*, 2005, **98**, 758.
25. K. Ishihara, N. Shibata, S. Tanaka, Y. Iwasaki, T. Kurosaki and N. Nakabayashi, *Journal of Biomedical Materials Research*, 1996, **32**, 401.
26. D. D'Souza, *http://radiopaedia.org,* 2009, **article ID: 5436**.
27. K. Wnuczko and M. Szczepański, *Pol. Merkuriusz Lek.*, 2007, **23**, 60.
28. R. F. Furchgott and J. V. Zawadzki, *Nature*, 1980, **288**, 373.
29. T. F. Lüscher and M. Barton, *Clin. Cardiol.*, 1997, **20**, 3.
30. J. R. Berrazueta, P. López-Jaramillo and S. Moncada, *Rev. Esp. Cardiol.*, 1990, **43**, 421.

31. V. A. Belitser, A. A. Musialkovskaia, T. N. Platonova and I. M. Ena, *Voprosy meditsinskoj khimii*, 1987, **33**, 8.
32. P. Carmeliet, *Nat. Med.*, 200, **6**, 389.
33. A. De Mell, G. Jell, M. M. Stevens and A. M. Seifalian, *Biomacromolecules*, 2008, **9**, 2969.
34. R. O. Hynes, *Cell*, 1992, **3**, 11.
35. G. M. Edelman, *Cell Commun. Adhes.*, 1993, **1**, 1.
36. A. E. Aplin, A. Howe, S. K. Alahari and R. L. Juliano, *Pharmacol. Rev.*, 1998, **50**, 197.
37. F. Rosso, A. Giordano, M. Barbarisi and A. Barbarisi, *J. Cell. Physiol.*, 2003, **199**, 174.
38. U. Hersel, C. Dahmen and H. Kessler, *Biomaterials*, 2003, **24**, 4385.
39. M. S. Perlin L., Rimmer S., *Soft Matter*, 2008, **4**, 2331.
40. P. H. Blit, Y. H. Shen, M. J. Ernsting, K. A. Woodhouse and J. P. Santerre, *J. Biomed. Mater. Res.*, 2010, **94**, 1226.
41. G. van der Pluijm, H. J. Vloedgraven, B. Ivanov, F. A. Robey, W. J. Grzesik, P. G. Robey, S. E. Papapoulos and C. W. Lowik, *Cancer Res.*, 1996, **56**, 1948.
42. R. McMillan, *J. Biomed. Mater. Res.*, 2001, **54**, 272.
43. W. J. Kao, *Biomaterials*, 1999, **20**, 2213.
44. G. Pande, *Curr. Opin. Cell Biol.*, 2000, **12**, 569.
45. Y. Wei, Y. Ji, L. Xiao, Q. Lin and J. Ji, *Colloids and Surfaces. B. Biointerfaces*, 2011, **84**, 369.
46. II. Ccylan, A. B. Tckinay and M. O. Gulcr, *Biomaterials*, 2011, **32**, 8797.
47. L. J. Taite, P. Yang, H. W. Jun and J. L. West, *Journal of Biomedial Materials Research Part B: Applied Biomaterials*, **84**, 108.
48. Y. Xiao and G. A. Truskey, *Biophysic Journal*, 1996, **71**, 2869.
49. E. Zamir and B. Geiger, *J. Cell Sci.*, 2001, **114**, 3583.
50. F. Re, *J. Cell Biol.*, 1994, **127**, 537.
51. D. G. Stupack, *J. Cell Biol.*, 2001, **155**, 459.
52. P. J. Keely, A. M. Fong, M. M. Zutter and S. A. Santoro, *J. Cell Sci.*, 1995, **108**, 595.
53. Z. Ma, C. Gao, Y. Gong, J. Ji and J. Shen, *J. Biomed. Mater. Res.*, 2002, **63**, 838.
54. M. Zuwei, M. Zhengwei and G. Changyou, *Colloids Surf. B. Biointerfaces*, 2007, **60**, 137.
55. L. Ji, E. Kang and K. Neoh, *Langmuir*, 2002, **18**, 9035.
56. G. I. Ignacio C., Oréfice R., *J. Appl. Polym. Sci.*, 2011, **121**, 3501.
57. B. Gupta, J. Hilborn, I. Bisson and P. Frey, *J. Appl. Polym. Sci.*, 2001, **81**, 2993.
58. G. Hermanson, ed., *Bioconjugate techniques.*, 2nd edn., Elsevier Inc., 2008.
59. *http://www.sigmaaldrich.com*.
60. N. Chansook and S. Kiatkamjornwong, *J. Appl. Polym. Sci.*, 2002, **89**, 1952.
61. M. H. Lee, K. J. Yoon and S.-W. Ko, *J. Appl. Polym. Sci.*, 2000, **78**, 1986.
62. A. K. Mohanty, S. Parija and M. Misra, *J. Appl. Polym. Sci.*, 1996, **60**, 931.
63. Y. M. Shin, Y. B. Lee, S. J. Kim and J. K. Kang *Biomacromolecules*, 2012, **13**, 2020.
64. C. Gao, J. Guan, Y. Zhu and J. Shen, *Macromol. Biosci.*, 2003, **3**, 157.
65. P. H. Blit, W. G. McClung, J. L. Brash, K. A. Woodhouse and J. P. Santerre, *Biomaterials*, 2011, **32**, 5790-5800.
66. Y. Lei, M. Remy, C. Labrugere and M.-C. Durrieu, *J. Mater. Sci. - Mater. Med.*, 2012, **23**, 2761.
67. Y. Weng, J. Chen and Q. Tu, *Interface Focus*, 2012, **2**, 356.
68. M. Sgarioto, P. Vigneron and J. Patterson, *C. R. Biol.*, 2012, **335**, 520-528.

THE USE OF CELL OUTER MEMBRANE MIMETIC SURFACES IN ORDER TO OBTAIN CLOT RESISTANT COATINGS

M. Trzaskowski*, B. Butruk and T. Ciach

Biomedical Engineering Laboratory, Faculty of Chemical and Process Engineering, Warsaw University of Technology, Poland
*m.trzaskowski@ichip.pw.edu.pl

1 INTRODUCTION

Polymers that are nowadays in use in construction of artificial cardiovascular implants need to be modified to meet current strict requirements in terms of biocompatibility and especially compatibility with blood. This is caused by the phenomenon of clotting of blood which happens on unnatural surfaces during a complex process called blood coagulation cascade.[1] Thus, every artificial device that contacts with blood in the body of patient poses a threat of spontaneous formation of clots.[2] Natural blood vessels, unlike artificial ones, have their own natural protection against spontaneous formation of blood clots. This protection is provided by endothelium cells, which are the cells that cover inner surfaces of all blood vessels.[3] Endothelium cells contact each other very closely creating a very homogenous layer of a clot-protecting coating which prevents adhesion of platelets, clotting factors and other small and large molecules on the walls of blood vessels.[4]

Numerous attempts have already been undertaken to modify surfaces of existing materials in order to provide necessary protection against adhesion of clot forming agents, very often by coating the materials with bioactive substances. In the last few years many of these studies involved the use of phospholipids in order to create coatings in form of cell outer membrane mimetic surfaces that would be acting like cell membranes of living organisms.

This paper is going to shortly describe the current studies of the use of phospholipids for clot-protecting materials and other biomedical applications. The second part of the paper will show the results of the authors' own attempt of creating a clot resistant material with the use of phospholipids.

2 PHOSPHOLIPIDS IN BIOMEDICAL ENGINEERING

2.1 Phospholipids

Phospholipids are a family of chemical compounds which all consist of phosphate combined with amine (e.g. choline, serine) and at least one carbohydrate chain. There are three main groups of phospholipids: glycerophospholipids, sphingolipids, and ether phospholipids, which differ from each other in the backbone molecule and types of bonds between functional groups of the molecule.[5] The most important and most numerous group are glycerophospholipids, which structures are based on the structure of glycerol. The fact that phospholipids consist of non-polar fatty acid chains as well as phosphate combined with amine – strongly polar groups, results in amphiphilic properties of these molecules. That means that they are built of hydrophilic and hydrophobic parts. For that reason phospholipids have natural ability to form micelles, vesicles and double membrane structures in water environment, and thus are used by cells of every living organism in construction of their cell membranes.[1]

2.2 Phosphatidylcholines

Phosphatidylcholines are a group of phospholipids which are the main component of human cell membranes. Structure of an example phosphatidylcholine molecule is shown in Figure 1. Polar part of the molecule marked in the figure is composed of phosphate and choline (phosphorylcholine group). Non-polar part of the molecule is build of two fatty acid chains. Specific compounds from the phosphatidylcholine family differ from each other only in lengths and unsaturation of fatty acid chains that build them.[5]

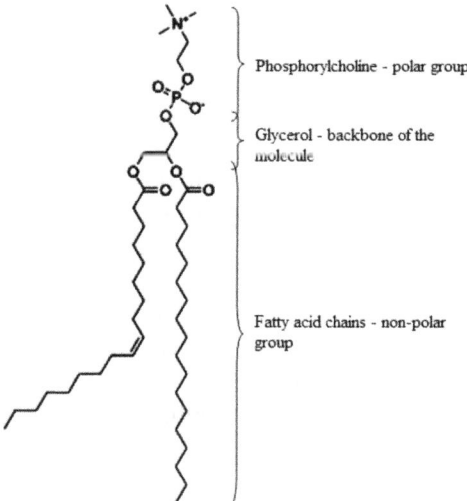

Phosphorylcholine - polar group

Glycerol - backbone of the molecule

Fatty acid chains - non-polar group

Figure 1 *Structure of a phosphatidylcholine molecule.*

Phosphatidylcholines (also known as lecithins) are present in the majority of living organisms. However, they are usually extracted from soya and rapeseed for industrial purposes. Egg yolk is also a reach source of lecithin.[6]

Phosphatidylcholines are widely used in the food industry. Their properties are used in production and stabilization of various emulsions. They are also used as food softeners. Lecithins are metabolized in human organisms without any side effects and thus they are safe food additives.[6,7]

Lecithins, apart from building cell membranes, also take part in many important metabolic processes. They are essential for proper work of nervous system as they build myelin sheaths.[8,9] Deficiency of phosphatidylcholines may cause memory problems.[10] Lecithins are also needed for proper functionality of liver, stomach, and cholesterol metabolism[11-13]. They assist in absorption of vitamins and counteract aging processes.[14] Because of big organism requirement lecithins are often consumed in form of dietary supplements.[6]

2.3 Antithrombogenic Properties of Phosphatidylcholines

Many research groups interest in phoshatidylcholines as compounds that can potentially improve properties of surfaces of biomaterials, especially materials used in construction of cardiovascular implants because of the antithrombogenic properties of phosphatidylcholine modified materials.[15,16] The reason of that interest is the fact that membranes of endothelium cells which contact with blood in natural conditions, are mainly built of phosphatidylcholines. Moreover phosphatidylcholines are able to spontaneously form structures similar to natural cell membranes called artificial cell membranes.

The mechanism of counteracting blood clotting by cell membrane mimetic structures is probably very complex and because of that – not yet fully understood. However, in the most of researchers' opinion the ability of bonding large amounts of water by the surface of the structure is very important in this phenomenon. One phosphorylcholine group can attract and hold up to 22 molecules of water thus protecting the surface of the structure from non-specific adsorption of proteins and blood cells.[15,17] Some scientists are also thinking that phosphorylcholine groups have an ability to selectively bond with proteins that prevent blood from clotting.[15]

Recently many attempts have been undertaken to connect phosphatidylcholines with various types of polymers in order to obtain materials covered with artificial cell membranes, resistant to formation of blood clots on their surfaces.[16]

2.4   Phosphatidylcholines in Biomaterial Engineering

In order to obtain clot-resistant surfaces many methods of connecting phosphatidylcholines with polymers have been applied. The majority of these methods is based on the use of a vinyl phospholipid monomer – 2-methacryloyloxyethyl phosphorylcholine (MPC) as a phosphatidylcholine particle (Figure 2).

One of the attempts was based on designing and fabricating polyurethane containing special segments with reactive functional groups (in this case OH-). This polymer was then used in reaction with MPC. As a result of this reaction, the surface of final polyurethane material has been covered with a layer of phospholipids. The prepared material exhibited very low adhesion of platelets.[18]

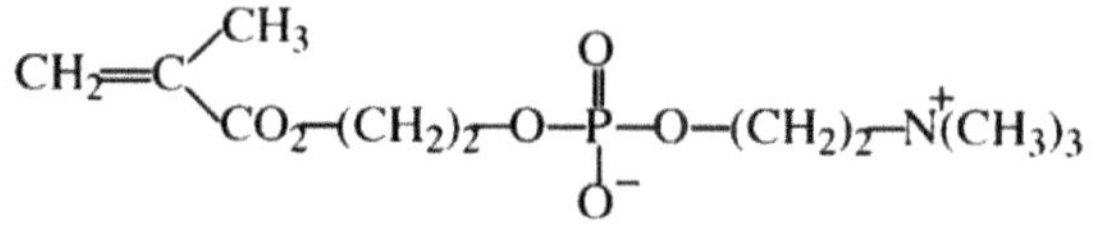

$$CH_2=C \underset{CO_2-(CH_2)_2-O-\overset{O}{\underset{O^-}{\overset{\|}{P}}}-O-(CH_2)_2-N^+(CH_3)_3}{\overset{CH_3}{}}$$

Figure 2 *Structure of 2-methacryloyloxyethyl phosphorylcholine.*

MPC was also used for preparation of segment polymers with the use of the reaction of free radical copolymerization with lauryl methacrylate. Elastic blood dialysis membranes made with such fabricated material were very compatible with blood and prevented microorganism adhesion on their surfaces.[19]

Another use of MPC was the modification of blood dialysis membranes made of cellulose acetate. In this case a copolymer of MPC and n-buthyl methacrylate has been mixed in proper ratio with cellulose acetate. Such prepared mixture has been used to prepare membranes. Also in this case results of blood clotting tests indicated very good properties of the material.[20]

Similar strategy was applied in order to fabricate clot resistant material for blood storage vessels. In this case a copolymer containing MPC was mixed with polysulphone.[21]

3 THE AIM OF THE STUDY

The aim of presented authors' own study was to create a phosphatidylcholine coating on the surface of polyurethane made of natural phosphatidylcholine (lecithin from soya beans) which would mimic a natural cell outer membrane. Such coating is expected to provide protection from clot formation as well as improve biocompatibility of the material that would be consequently more suitable to use for construction of artificial blood vessels.

4 MATERIALS AND METHODS

4.1 Materials

In the described study following materials were used. Cyclohexanone supplied by Carlo Erba, lecithin from soya beans (Merck) – as a source of phosphatidylcholine. Base material, on which coatings were produced were ether-based polyurethane foil Walopur 4201 AU supplied by Epurex Films and medical grade ChronoThane polyurethane supplied by CardioTech. Polyurethane used in preparation of coatings was Estane 5715P supplied by Lubrizol.

4.2 Methods

The fabrication of phosphatidylcholine coatings was performed with the use of a dip-coating technique.

Firstly, a 1% solution of polyurethane in cyclohexanone, was made. The solution was then enriched with a certain amount of phosphatidylcholine (0.5-2% m/v) and stirred continuously until the total dissolution of all components.

In the next step samples in form of square pieces of polyurethane foil, previously cleaned by washing in ethanol, were immersed in the solution for a few seconds. Then the samples were dried for a few minutes in order to let the solvent evaporate from the surface of the sample.

After this step samples were immersed in distilled water for 10 minutes. The purpose of the water treatment was not only to rinse out the remaining solvent and unbound particles but also to enable the surface phosphatidylcholine particles to reorganize by turning their polar sides to the outside and thus obtain the structure of cell outer membrane (Figure 3). The samples were then thoroughly dried and examined.

4.3 Methods of Examination of Obtained Materials

The analysis of the chemical composition of the surface layer of the fabricated materials was performed with the FT-IR/ATR technique. Spectra of the samples were recorded using Thermo Scientific Nicolet 6700 FT-IR spectrometer with ATR pickup.

The change of the contact angle of the surface of samples after the modification process was measured with the use of KSV CAM 200 goniometer and Attension Theta software supplied by Biolin Scientific.

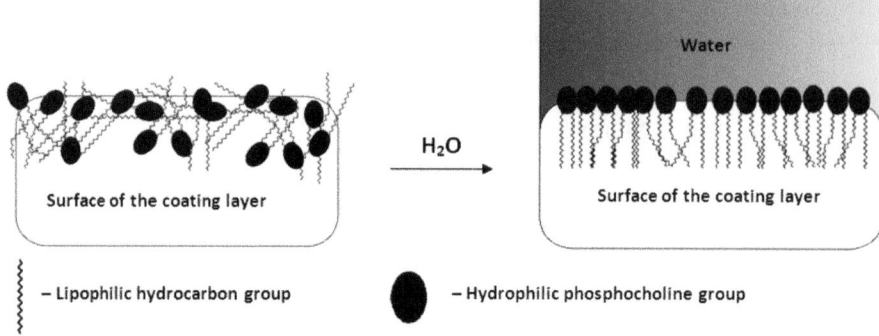

Figure 3   *The process of water treatment in order to organize phosphatidylcholine particles.*

Cytotoxicity of obtained materials, was measured with the use of MTT method on L929 mouse fibroblast cell line, after 24 hours of exposure to examined materials.

Blood compatibility was examined by courtesy of employees of the Jagiellonian University Medical College, Chair of Pharmacology in Cracow, Poland. The analysis was performed with the use of *Impact-R* method and plate-cone analyzer (CPA *Impact-R* DiaMed AG). The procedure for blood compatibility analysis was performed as follows. Examined materials were placed on the plate of the analyzer, which afterwards started to rotate. A specific amount of blood (130 µl of human blood, from a healthy volunteer, right after donation) was dropped on the center of rotating sample. The rotating movement (720 rpm, 300 s) was causing the blood to flow on the surface of the sample thus imitating a natural flow of the blood in a vessel. The blood was afterwards collected and examined for the platelet-material interactions.

Samples after the blood compatibility tests were examined with the use of SEM microscope.

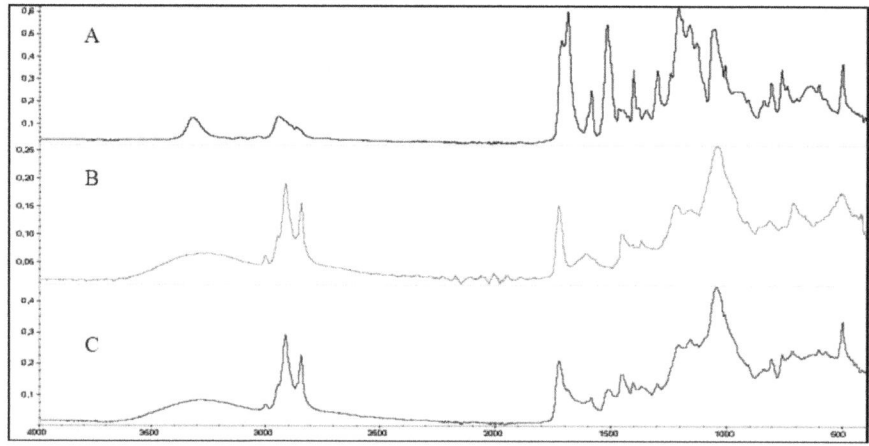

Figure 4   *The comparison of FTIR-ATR spectra of unmodified polyurethane (A), pure phosphatidylcholine (B) and a sample after the modification process (C).*

5 RESULTS AND DISCUSSION

5.1 FT-IR/ATR Analysis

FT-IR/ATR analysis (Figure 4) confirmed that the surface of the material is covered with phosphatidylcholine. This was proved by the appearance of new signals in the spectrum of the sample after modification (two sharp signals at approximately 2750 and 2950 and a broad signal at 3500-3000 wave number). These signals come from CH_2 groups from hydrocarbon chains and amine group of the phosphatidylcholine.

5.2 Contact Angle Analysis

Contact angle examination showed a significant difference between samples before and after the modification process (Figure 5). The material became strongly hydrophilic which not only proves that phosphatidylcholine is present on the surface but also suggests that it

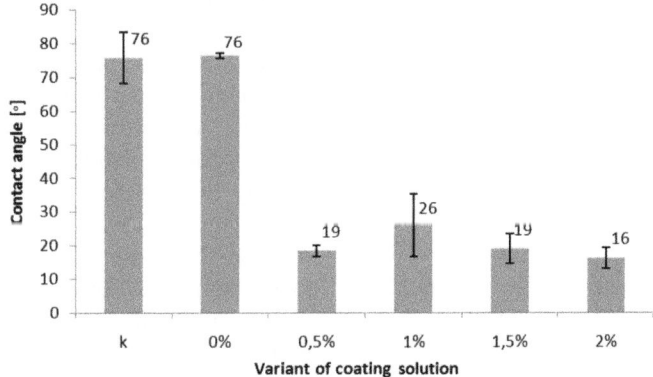

Figure 5   *Results of contact angle measurements for pure polyurethane (control) and different variants of fabricated material (differing in concentration of PC solution "k" stands for pure polyurethane).*

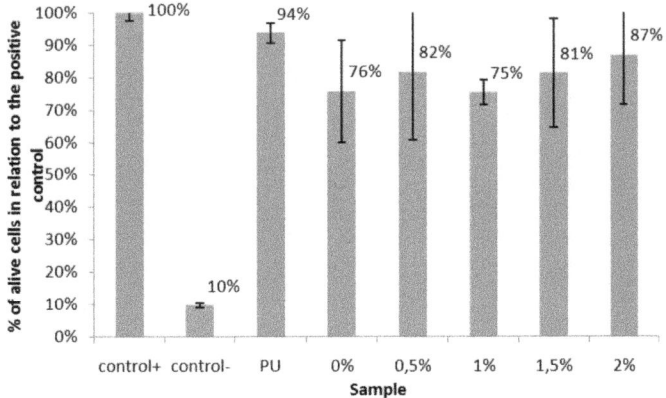

Figure 6   *The results of MTT viability test done after 24 hour exposure of cells to examined materials. "control+" stands for cells cultivated without any material exposure, "control-" – cells cultivated without material exposure with triton X added to the medium. Series contain of 3 replicates.*

is oriented with its polar sides to the outside. Every presented series of results contains of minimum 5 measurements for every of 3 samples of each type.

5.3 Cytotoxicity Analysis

MTT test done on cells after the 24 hour material exposure did not indicate significant difference in viability between material before and after the modification process. Materials are considered non-cytotoxic when the MTT test results are no worse than 70% when compared to the control. This proves that the obtained material was not cytotoxic (Figure 6).

5.4 Blood compatibility analysis

The analysis of number of platelets present in blood after the Impact-R test is an indirect method of determining how many platelets adhered to the surface of examined material. The results indicate that materials with phosphatidylcholine coatings cause very small adhesion of platelets. (Figure 7).

The analysis of number of platelet aggregates in blood after the contact with material was performed to determine the influence of material on the activation of platelets. The more aggregates in population of platelets the worse material properties. The results show that all phosphatidylcholine coated materials exhibit smaller platelet activation than the base material, and the 1.5% coating performs similarly to the reference antithrombogenic material Figure 8).

SEM images of samples after blood incubation tests show much less blood cells on the surface of the materials after the modification process, than on the base material (Figure 9). This suggests that the final material has desired anti-thrombogenic properties.

Holes that are visible on the surface of materials are probably increasing the adhesion of blood cells and are dependent on phosphatidylcholine concentration in the modifying

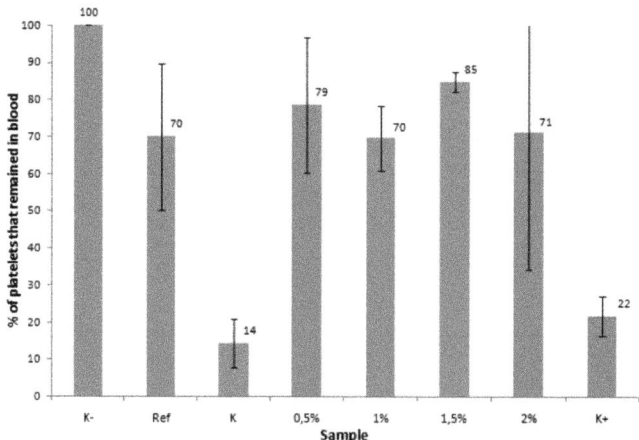

Figure 7   *The results of number of platelets analysis. Higher number of platelets in blood after the test indicates better antithrombogenic properties of the examined material. K- stands for base blood (negative control), Ref for reference antithrombogenic material, K is material before the modification process, and K+ is blood activated with ADP (positive control). All series contain of 3 replicates.*

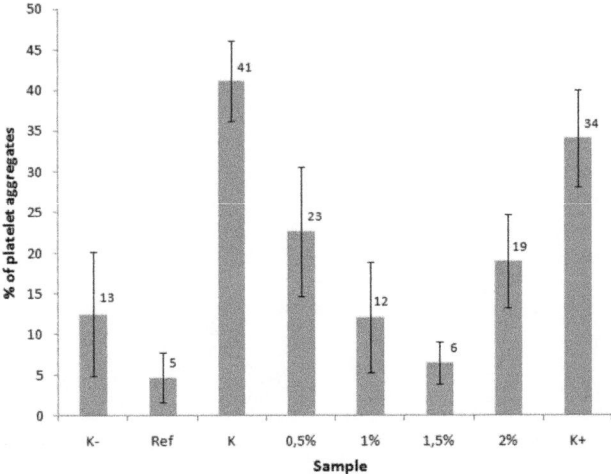

Figure 8 *The results of number of platelet aggregates in the population of platelets analysis. Higher number of platelet aggregates in blood after the test indicates worse antithrombogenic properties of the examined material. K- stands for base blood (negative control), Ref for reference antithrombogenic material, K is material before the modification process, and K+ is blood activated with ADP (positive control). All series contain of 3 replicates.*

solution. This is why SEM images also helped in determining the optimum phosphatidylcholine concentration to be 1.5%.

6 CONCLUSION

A simple method of coating polyurethane with phosphatidylcholine was presented. It was confirmed by analytical methods that the procedure of coating fabrication was successful. The results of cytotoxicity and blood compatibility tests show that the fabricated materials exhibit antithrombogenic properties and are safe for mammalian cells. The research shows that the coating made with the use of natural phosphatidylcholine can be as effective as coatings made of phospholipid monomer (MPC) and potentially useful in cardiovascular implants construction.

Acknowledgements

The presented work has been supported by the European Union in the framework of European Social Fund through the Warsaw University of Technology Development Program.

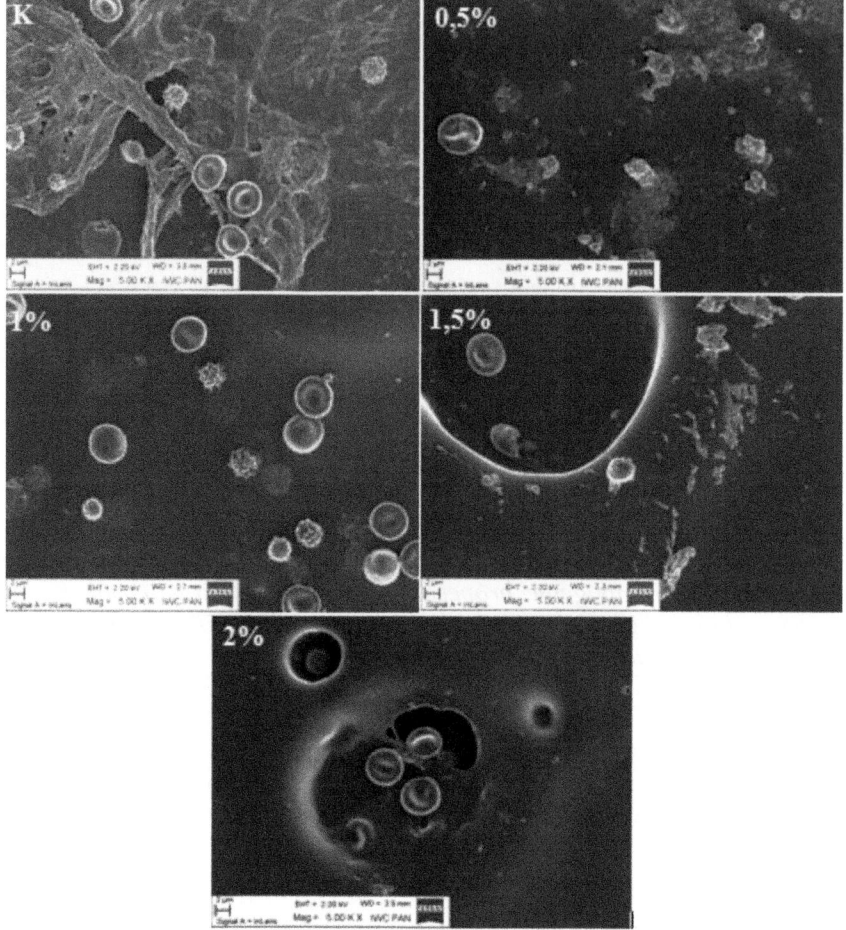

Figure 9   *SEM images of examined materials after blood incubation tests "K" stands for pure polyurethane (magnification 1000x).*

References

1. L. Stryer, *Biochemistry*, Wydawnictwo Naukowe PWN, Warszawa, 4th edn., 2003.
2. A. E. Aksoy, V. Hasirci and N. Hasirci, *Macromol. Symp.*, 2008, **269**, 145-153.
3. B. A. Butruk, P. A. Ziêtek and T. Ciach, *Cent. Eur. J. Chem.*, 2011, **9**, 1039-1045.
4. A. de Mel, G. Jell, M. M. Stevens and A. M. Seifalian, *Biomacromolecules*, 2008, **9**, 2969-2979.
5. X. X. Zheng G., *Biotechnol. Adv.*, 2005, **23**, 203-259.
6. B. F. Szuhaj, *J. Am. Oil Chem. Soc.*, 1982, **60**, 306-309.
7. I. Kralova and J. Sjöblom, *J. Dispersion Sci. Technol.*, 2010, **31**, 702-702.
8. K. Mishima, T. Ogihara, M. Tomita and K. Satoh, *Chem. Phys. Lipids*, 1992, **62**, 87-91.
9. A. H. Ousley and P. Morell, *J. Biol. Chem.*, 1992, **267**, 10362-10369.
10. W. W. Li and L. N. Ren, *Journal of Dalian Medical University*, 2010, **32**, 509-511, 515.

11. S. Demirbilek, I. Gürses, N. Sezgin, A. Karaman and N. Gürbüz, *J. Pediatr. Surg.*, 2004, **39**, 57-62.
12. M. C. Phillips, *Hepatology*, 1990, **12**, 75S-82S.
13. R. M. Epand, R. F. Epand, D. W. Hughes, B. G. Sayer, N. Borochov, D. Bach and E. Wachtel, *Chem. Phys. Lipids*, 2005, **135**, 39-53.
14. B. Liu, J. Du, J. Zeng, C. Chen and S. Niu, *Eur. Food Res. Technol.*, 2009, **230**, 325-331.
15. K. G. Marra, T. M. Winger, S. R. Hanson and E. L. Chaikof, *Macromolecules*, 1997, **30**, 6483-6488.
16. H. Kyun Kim, K. Kim and Y. Byun, *Biomaterials*, 2005, **26**, 3435-3444.
17. M. M. Zong and Y. K. Gong, *Chin. J. Polym. Sci.*, 2011, **29**, 53-64.
18. A. Korematsu, Y. Takemoto, T. Nakaya and H. Inoue, *Biomaterials*, 2002, **23**, 263-271.
19. A. L. Lewis, Z. L. Cumming, H. H. Goreish, L. C. Kirkwood, L. A. Tolhurst and P. W. Stratford, *Biomaterials*, 2001, **22**, 99-111.
20. S. H. Ye, J. Watanabe, Y. Iwasaki and K. Ishihara, *Biomaterials*, 2003, **24**, 4143-4152.
21. T. Hasegawa, Y. Iwasaki and K. Ishihara, *Biomaterials*, 2001, **22**, 243-251.

CONFERENCE OUTLOOK

1ST INTERNATIONAL CONFERENCE ON BIOLOGICAL AND BIOMIMETIC ADHESIVES

R. Santos,* M. Almeida, M. Lopes and L.P. Lopes

Biomedical and Oral Sciences Research Unit (UICOB), University of Lisbon, Portugal.
*romana.santos@campus.ul.pt

In the course of the second year of COST Action TD0906 "Biological adhesives: from biology to biomimetics (http://www.cost-bioadhesives.org)", the 1st International Conference on Biological and Biomimetic Adhesives was organised by the Management Committee Vice-Chair, Romana Santos, and her team at the School of Dentistry of the University of Lisbon, Portugal. The event spanned from May 9th to 11th 2012 (Figure1), with the aim of bringing together researchers and manufacturers in the forefront of adhesion science and technology.

The inspiration for this meeting was the broad and increasing interest in development of biomimetic adhesives from both the scientific and industrial communities. In fact, adhesives found in nature combine many inspiring properties such as sophisticated hierarchical organization, versatility and adaptability, which are rarely met by commercially available synthetic adhesives. Thus, it is necessary to elucidate the basic

Figure 1 *Logo of the 1st International Conference on Biological and Biomimetic Adhesives held at the School of Dentistry of the University of Lisbon in May 2012, showing the Lisbon bridge over the Tagus river (R. Santos).*

Figure 2   *A group photo of the conference participants.*

components and building principles selected by evolution in order to propose more reliable, efficient and environmentally-friendly materials. The potential of the biomimetic approach in the field of adhesives remains far from fully exploited. Nature offers a vast repository of inspiration in the form of organisms utilizing bioadhesives for a variety of functions. The challenge is to understand the mode of action of these systems, their unifying themes and function-specific adaptations, and to transform this knowledge into technology by developing new synthetic biomimetic adhesives with improved performance.

Altogether 98 participants from 24 countries worldwide (Figure 2), including the USA, Japan and New Zealand, presented their most recent research on fundamental or applied adhesion and related topics. More than 60 oral and poster presentations, including 10 world-renowned plenary speakers and 1 mini-workshop on patents by UL-INOVAR, were presented over a two and a half day period. The programme provided a complete overview of the chemical and structural characterization of biological adhesives (sessions I and II), the mechanical testing of adhesives and theory (session III) as well as the fabrication and application of biomimetic adhesives (sessions IV and V).

The list of participants in presented in Table 1. The detailed programme as well as the abstract book of the 1st International Conference on Biological and Biomimetic Adhesives are available at https://sites.google.com/site/1sticbbaa/home. The conference was a further success in terms of gender balance (62% male and 38% female participants), and in attracting Early Stage Researchers (36%). In addition, the participants' affiliations reflected the widespread interest in bioadhesion and its applications, with researchers participating from more than 40 different research institutes and 8 companies.

Acknowledgements

General support for the 1st International Conference on Biological and Biomimetic Adhesives was provided by Faculdade de Medicina Dentária da Universidade de Lisboa, Reitoria da Universidade de Lisboa, Sociedade Portuguesa de Estomatologia e Medicina Dentária, Fundação Luso-Americana and COST Action TD0906. We are grateful to the following plenary and invited speakers that presented their work: Alexandre Cavalheiro, Andrew Smith, Costantino Creton, Julius Vancso, Kathryn Wahl, Kei Kamino, Lothar Schlosser, Manoj Chaudhury, Metin Sitti Nuno Silva and Russell Stewart. We would also like to thank all participants and members of the scientific committee as well as the session chairs for all your efforts and all the contributions that lead to a successful conference. Final thanks goes to all members of the local organizing, logistics and technical support teams in Lisbon.

Table 1 *List of conference participants (by alphabetic order)*

Local Organizing Committee

Almeida Marise	Portugal	malmeida2@campus.ul.pt
Lopes Manuela	Portugal	manuela.lopes@fmd.ul.pt
Madeira Cidália	Portugal	cidalia.madeira@fmd.ul.pt
Pires Lopes Luis	Portugal	pires.lopes@fmd.ul.pt
Santos Romana	Portugal	romana.santos@campus.ul.pt

Keynote speakers

Cavalheiro Alexandre	Portugal	acavalheiro@mac.com
Chaudhury Manoj	USA	mkc4@lehigh.edu
Creton Costantino	France	costantino.creton@espci.fr
Kamino Kei	Japan	kamino-kei@nite.go.jp
Schlosser Lothar	Switzerland	lothar.schloesser@geistlich.ch
Sitti Metin	USA	sitti@cmu.edu
Smith Andrew	USA	asmith@ithaca.edu
Stewart Russell	USA	rstewart@eng.utah.edu
Vancso Julius	The Netherlands	g.j.vancso@utwente.nl
Wahl Kathryn	USA	kathryn.wahl@nrl.navy.mil

Participants

Akerboom Sabine	The Netherlands	sabine.akerboom@wur.nl
Aldred Nicholas	United Kingdom	nicholas.aldred@ncl.ac.uk
Angarano Marco	Germany	marco.angarano@fmf.uni-freiburg.de
Athanassiadou Eleftheria	Greece	eathan@ari.gr
Barnes Jon	United Kingdom	Jon.Barnes@glasgow.ac.uk
Bennemann Michael	Germany	bennemann@bio2.rwth-aachen.de
Berglin Mattias	Sweden	mattias.berglin@cmb.gu.se
Birkedal Henrik	Denmark	hbirkedal@chem.au.dk
Bohn Holger	Germany	holger.bohn@biologie.uni-freiburg.de
Bras Thomas	Belgium	thomas.bras@umons.ac.be
Braun Julius	Germany	julius.braun@webdays.de
Bundschuh Sven	Germany	sven.bundschuh@kit.edu

Ciach Tomasz	Poland	t.ciach@ichip.pw.edu.pl
Coimbra Carla	Portugal	cccarlacoimbra@gmail.com
Cui Jiaxi	Germany	cui@mpip-mainz.mpg.de
Cyran Norbert	Austria	nbc555@gmx.net
del Campo Aranzazu	Germany	delcampo@mpip-mainz.mpg.de
Demeuldre Mélanie	Belgium	melanie.demeuldre@umons.ac.be
Di Fino Alessio	United Kingdom	alessio.di-fino@newcastle.ac.uk
Dimartino Simone	New Zealand	simone.dimartino@canterbury.ac.nz
Ditsche-Kuru Petra	Germany	pditschekuru@zoologie.uni-kiel.de
Dudek Katarzyna	Poland	katarzyna.dudek@synthosgroup.com
Endlein Thomas	United Kingdom	thomas.endlein@glasgow.ac.uk
Flammang Patrick	Belgium	Patrick.Flammang@umons.ac.be
Foreman Paul	USA	paul.foreman@us.henkel.com
Frenzke Lena	Germany	lena.frenzke@mailbox.tu-dresden.de
Furtos Gabriel	Romania	gfurtos@yahoo.co.uk
Fusi Paola	Italy	paola.fusi@unimib.it
Gorb Stanislav	Germany	sgorb@zoologie.uni-kiel.de
Greco Giuliano	Italy	greco.giuliano@gmail.com
Grunér Mathias	Finland	Mathias.Gruner@vtt.fi
Guler Mustafa	Turkey	moguler@unam.bilkent.edu.tr
Haber Meir	Israel	meir@algawish.com
Hennebert Elise	Belgium	elise.hennebert@umons.ac.be
Higgins Laila	Ireland	laila.higgins@ucdconnect.ie
Hoffmann Michael	Germany	michael.hoffmann@ifam.fraunhofer.de
Hosoda Inoue Naoe	Japan	Hosoda.Naoe@nims.go.jp
Ishii Naoki	Japan	Naoki_Ishii@terumo.co.jp
Jonker Jaimie-Leigh	Ireland	jaimieleigh.jonker@gmail.com
Kamperman Marleen	The Netherlands	marleen.kamperman@wur.nl
Kasak Peter	Slovakia	peter.kasak@savba.sk
Kaucic Venceslav	Slovenia	kaucic@ki.si
Kiwi Juan	Switzerland	john.kiwi@epfl.ch
Labonte David	United Kingdom	dl416@cam.ac.uk
Ladurner Peter	Austria	peter.ladurner@uibk.ac.at
Lengerer Birgit	Austria	birgit.lengerer@student.uibk.ac.at
Linder Markus	Finland	markus.linder@vtt.fi
Mano João	Portugal	jmano@dep.uminho.pt
Mazzolai Barbara	Italy	barbara.mazzolai@iit.it
Meric Zeynep	Germany	zzeynepmeric@gmail.com
Mostaert Anika	Ireland	anika.mostaert@ucd.ie
Neto Ana Isabel	Portugal	isabel.neto@dep.uminho.pt
Ngo Thi Chinh	Belgium	thichinh.ngo@umons.ac.be
Nürnberger Sylvia	Austria	Sylvia.Nuernberger@meduniwien.ac.at
Palacios Maria Jose	Spain	mjescoto@inescop.es
Power Anne Marie	Ireland	annemarie.power@nuigalway.ie
Rischka Klaus	Germany	klaus.rischka@ifam.fraunhofer.de
Robinson Adam	United Kingdom	ar433@cam.ac.uk
Röhrig Michael	Germany	michael.roehrig@kit.edu
Samuel Diana	United Kingdom	d.samuel.1@research.gla.ac.uk
Schiffels Peter	Germany	peter.schiffels@ifam.fraunhofer.de
Schmitt Christian	Germany	christian.schmitt@uni-tuebingen.de

Schiffels Peter	Germany	peter.schiffels@ifam.fraunhofer.de
Schmitt Christian	Germany	christian.schmitt@uni-tuebingen.de
Schwotzer Willi	Switzerland	willi.schwotzer@nolax.com
Serrano Cristina	Spain	serrano@mpip-mainz.mpg.de
Shafiq Zahid	Germany	shafiq@mpip-mainz.mpg.de
Silva Nuno	Portugal	inovar@campus.ul.pt
Sobolciak Patrik	Slovakia	Patrik.Sobolciak@savba.sk
Stanuch Wojciech	Poland	wojciech.stanuch@synthosgroup.com
Tawheed Mohamed	Germany	Dr_tawheed@yahoo.com
Tekinay Ayse	Turkey	atekinay@bilkent.edu.tr
Tomer Guy	Israel	tomer@life-bond.com
Tomoaia-Cotisel Maria	Romania	mcotisel.chem.ubbcluj.ro@gmail.com
Tramacere Francesca	Italy	francesca.tramacere@iit.it
Trzaskowski Maciej	Poland	m.trzaskowski@ichip.pw.edu.pl
Tseng Yao Hung	Denmark	yaohung.tseng@gmail.com
Van de Weerdt Cécile	Belgium	C.Vandeweerdt@ulg.ac.be
Voigt Dagmar	Germany	dagmarvoigt@web.de
Wyrwa Ralf	Germany	rw1@innovent-jena.de
Zietek Paulina	Poland	p.zietek@ichip.pw.edu.pl

Subject Index